CONNECTING BRAINS AND SOCIETY

CONNECTING BRAINS AND SOCIETY

The Present and Future of Brain Science

what is possible, what is desirable?

International Workshop
22 and 23 April 2004
Amsterdam
The Netherlands

Proceedings and Synthesis report

King Baudouin Foundation

Rathenau instituut

Colophon

Connecting Brains and Society
The Present and Future of Brain Science: what is possible, what is desirable?
International Workshop, 22 and 23 April 2004, Amsterdam, The Netherlands
Proceedings and Synthesis report

THIS IS A JOINT PUBLICATION OF
the King Baudouin Foundation (Brussels, Belgium) and
the Rathenau Institute (The Hague, The Netherlands)

WRITERS AND EDITORS
Peter Raeymaekers • Ph.D. Molecular Biology, science writer
Karin Rondia • Medical Doctor, science journalist
Marjan Slob • MA Philosophy, publicist

PROJECT MANAGERS
Sara Heesterbeek • Rathenau Institute
Rinie van Est • Rathenau Institute
Ann Nicoletti • King Baudouin Foundation
Tinne Vandensande • King Baudouin Foundation

TRANSLATION AND LANGUAGE REVISION
Barry Brennan, Kings Translation & Copywriting
Linguanet, Brussels
Roger Thurman, English Text Company, The Hague

PHOTOGRAPHY
Sara Heesterbeek, p. 107

LAYOUT
Megaluna+Triumviraat, Brussels

PRINTING
New Goff

ISBN: 90-5130-478-1
Legal Depot: D/2004/2893/27

This publication is available free of charge:
On-line via www.kbs-frb.be or www.rathenau.nl
By e-mailing publi@kbs-frb.be or bestellingen@rathenau.nl
Or by phoning the King Baudouin Foundation office
Tel: +32-70-233 728 Fax: +32-70-233 727

With the support of the Belgian National Lottery

December 2004

Table of Contents

| Foreword |

Recent years have shown a strong increase in interest in the ethical and societal aspects of the brain sciences. One landmark event was the conference *Neuroethics: Mapping the Field*, which was organised by the Dana Foundation in San Francisco in May 2002. We would like to see the workshop 'Connecting Brains and Society: The Present and Future of Brain Sciences: what is possible, what is desirable?', held in April 2004 in Amsterdam, the Netherlands, as the European follow-up to the San Francisco conference.

The aim of the Amsterdam workshop was to give an overview of the current status of and expectations for the brain sciences and their related ethical and social issues. It initiated dialogue between natural and social scientists on the one hand and stakeholders on the other in regard to these topics. The workshop also served as a kick-off meeting for the ambitious European Citizens' Deliberation on Brain Science (the ECD project), which by means of citizen's panels seeks to involve citizens from nine European countries in the debate on societal issues raised by developments in the brain sciences.

The workshop in Amsterdam was sponsored and organised by the co-ordinator of the ECD project, the King Baudouin Foundation, in collaboration with the Rathenau Institute, which is the Dutch national Technology Assessment organisation and part of the Royal Netherlands Academy of Sciences. One reason why the King Baudouin Foundation asked the Rathenau Institute to co-organise and host the 'Connecting Brains and Society' workshop was that the Institute had been engaged in the study of the societal issues related to the brain sciences since the beginning of this millennium.

At that time, developments in the field of neurosciences, biotechnology, and information technology had led to the expectation that many new technologies for influencing human behaviour would become available in the mid-term. The Rathenau Institute used the term 'psychotechnology' to pack together these various developments. The initial focus was on using medicines to intervene in behaviour. The report *Pills and Psyche* (2002) outlined the development of (new) anti-depressants for adults and Ritalin for hyperactive children. Two years later it was followed by a book, *An Other Self* written by the philosopher Marjan Slob. This book describes a range of technologies, from pharmaceuticals and neuro-surgery to neuro-imaging and genetics, that may alter our understanding of our brains and may even be used to intervene directly in the human personality. *An Other Self* also reflects on the social and ethical questions that are related to these developments.

Since the beginning of this century, the term brain sciences has become an established expression and has come to the attention of social scientists, journalists, science museums and technology assessment and foresight organisations. Cognitive sciences have also come to be acknowledged as a key technology alongside nanotechnology, biotechnology and information technology. The idea that these technologies are converging and strengthening one another has come to the fore under the abbreviation NBIC as a new paradigm for looking at the development of the natural sciences. It is important that society be informed about and engaged with these developments in science from the onset. Of the four named key technologies, the brain sciences is the one that has received the least public attention. So, this seems to be the right time to get the public involved in the field of brain sciences, which, like no other scientific field influences the way we look at our fellow humans and ourselves. The European Citizens' Deliberation on Brain Science project is precisely trying to promote this.

The Rathenau Institute is pleased to have been able to host the workshop 'Connecting Brains and Society' and would like to thank all participants for their contributions. It would also like to thank the people who organised it within the space of just three months: Marjan Slob, Karin Ronda, and Peter Raeymaekers (the science journalists who besides organising, also mediated the workshop and wrote the proceedings), Gerrit Rauws, Tinne Vandensande and Ann Nicoletti (King Baudouin Foundation), and Sara Heesterbeek, Corine Wildschut and Rinie van Est (Rathenau Institute). We would like to continue our pleasant co-operation in the future, and hope that the results of the whole ECD project will be as relevant and inspiring as those of the Amsterdam workshop.

Jan Staman
Director Rathenau Institute

| Introduction |

'Connecting Brains and Society' is the first event of the ambitious European Citizens' Deliberation on Brain Science (ECD) initiative. Its overall objective is to design and implement a new way of interactive governance at European level. It aims to do so by making a concrete and tangible contribution to the public deliberation on brain sciences, research & development and related ethical and socio-political issues.

The ECD initiative will engage citizens from different European national and cultural contexts in sharing their perspectives, values and reasoning concerning the societal consequences of brain research, in identifying their differences as well as their commonalities, and in deliberating these issues with relevant experts, stakeholders and policy-makers. Finally, the citizens will address their recommendations to the scientific and research community at the European level and to stakeholders and policy-makers at the European, national and transnational level. As such, the initiative aims to make relevant inputs into European policy-making and the wider public debate. This innovative project is scheduled to run from 2004 to 2006.

A consortium of twelve partner organisations from nine different countries is involved, including parliamentary technology assessment bodies, science museums, academic institutions and public foundations with substantial experience of public participation, governance and communication relating to science, technology and society.

The King Baudouin Foundation initiated the ECD, in accordance with its Governance Programme and its ambition to stimulate the development of debate and decision-making models tailored to the transnational level. The Foundation wishes to create a stimulating environment in which individuals and institutions have access to new forms of social debate, decision-making processes and political approaches. The challenge is to give citizens a key role and encourage them to become involved in such issues based on their own questions, concerns and capabilities. Although, in the future the national and local levels will still be the key levels for harmonising governance processes, social challenges are becoming increasingly transnational in nature. In the coming years, there will be an even more acute need for debate and decision-making models tailored to the transnational level.

The growing Europeanisation of politics, economic globalisation and the need for new transnational governance practices warrant the choice of a European approach to the brain sciences.

The European Commission's plans for the European Research Area envisage greater interaction between scientists and various stakeholders (including citizens) in research and development and related policy-making (*Towards A European Research Area*. European Commission, 2001). The Science and Society Action Plan represents the European Commission's practical implementation of this policy (*Science and Society Action Plan*. European Commission, 2002). With its objectives, the ECD initiative wishes to meet EC calls for greater public involvement in the discussion of future research, technological decision-making and governance.

Before involving citizens in the discussions on the brain sciences, the European partner consortium of ECD decided to organise an international expert stakeholders' workshop, in order to investigate the technological state of affairs, the range of potential social consequences and the way they interact.
Both the workshop and the book publishing the full report are a collaborative initiative of the Rathenau Institute and the King Baudouin Foundation.

The workshop took place on 22 and 23 April 2004 in the Shaffy theatre of the prestigious Felix Meritis building in Amsterdam, a building that has always symbolised the commitment of the citizens of the city to the advancement of the arts and sciences. A select group of 25 European top-level scientists, cultural thinkers and stakeholders, all working at the cutting edge of the brain sciences, were invited to give their views on developments in the brain sciences from their own perspectives. Among them were physicians, neurologists, psychiatrists, cognitive and social scientists, philosophers, artists and representatives of stakeholder organisations (see appendix 2). The 'audience' was a group of representatives from the ECD consortium (see appendix 3). More information about the workshop set-up and the two-day programme can be found under the heading 'Aims and expectations' from page 19 onwards.

This book is divided in two main parts. Part I is devoted to the proceedings and faithfully follows the different sessions and the statements made by the experts and stakeholders. Part II consists of a synthesis report giving a more extensive overview of the issues which cropped up during the workshop, in a form which aims to be more coherent than what a lively workshop pressed for time can be by its nature. It also adds relevant quotes and insights from other specialists in the field.

A variety of opinions reflecting the different professional, cultural and national backgrounds of the participants surfaced during the workshop. This variety of viewpoints showed that the topic of the brain sciences has an enormous potential for citizens' deliberation; however it was also clear that the issues must be presented to citizens in appropriate 'language'.

Everyone agreed that the 'science of the brain' will increasingly provide the vocabulary with which we discuss ourselves. This puts pressure on traditional moral concepts like responsibility and free will. With techniques developed by the brain sciences, we shall also become able to 'improve' the emotional and cognitive states which we still find normal.

The King Baudouin Foundation wishes to thank all participants who were present at the Amsterdam workshop for their excellent work. They will remain a core group of stakeholders during the whole ECD participatory trajectory. The Foundation hopes that reading this book will reveal that 'mind matters'.

The time has come for the public, researchers and policy-makers in Europe to further explore the issues at stake in the field of the brain sciences.

Gerrit Rauws
Director of the Governance Programme
King Baudouin Foundation

| Executive Summary |

"Brain sciences cover not only the treatment of neurological disease, but also constitute an important narrative about what it is to be human", said Danish medical anthropologist Andreas Roepstorff at the European workshop on *The Present and Future of Brain Science: what is possible, what is desirable?* of which this book represents the proceedings. His statement nicely summarises how important it is for citizens to have an opinion on brain sciences. Brain diseases are forecast to become a large and growing burden for ageing Europeans, with predictions that by 2030 about 50 percent of the population will have a brain disease. And since 'braintalk' is becoming a dominant source of concepts by which we understand and articulate our humanity, our view of ourselves as human beings is at stake too.

The explicit aim of the Amsterdam workshop was for scientists and stakeholders working at the frontier of brain science to formulate what they consider to be the most pressing issues for citizens to debate. The workshop is the first step in an ambitious project, in which citizens will be encouraged to articulate their concerns and recommendations regarding developments in the brain sciences. During the first day of the workshop, seven experts presented the latest developments in their fields. The second day saw stakeholders and social scientists give short statements on what they consider to be the most pressing issues for citizens in Europe to debate; attendees then discussed the three most pressing topics in depth. All these presentations, and the lively debates which followed them, found a place in the first part of this book; the proceedings of the workshop. The second part offers a synthesis; it maps key social issues stirred up by the brain sciences.

Medicalisation
In the brain sciences there is a particularly thin line between real pathology and a medicalised social problem. The trend is to medically label conditions that are not necessarily pathological. Such 'medicalisation' especially comes into play for people with mood or behavioural problems. The rephrasing of the problems in medical terms can repress other, more socially oriented 'explanations' of them - and their accompanying treatments. This is something for the citizen to be alert to.

Enhancement and social justice
Brain sciences concern a variety of technologies which aim to cure neurological diseases. But often these methods - notably psychotropics - can also be used to enhance the cognitive functions, or the moods, of people who are not ill.

Many workshop attendees made a plea for proactive reflection on the societal impact of, specifically, cognitive enhancement. Cognitive enhancement will cause society to change incrementally. If the level of 'normal' cognitive performance rises, but only one section of the population has the resources to attain this new performance level, then this would reinforce social pressures and set new societal norms.

Brain technology, writ large, might also reduce the acceptability of being a certain type of person. If personality traits and behavioural characteristics become detectable on brain images, and optionally treatable, people may be tempted to discriminate against the bearer of those traits. This could in turn lead to a less tolerant and very normative society where exceptions to the norm could be considered deviant and the choice would be 'to be treated or re-educated'.

The dilemma between individual benefits and societal values is probably the core of the dilemma of brain-based enhancement. Enhancement is neither right nor wrong, but citizens would be advised to consider a series of questions as to safety use, freedom of choice, use in children, prevention of coercion by employers or schools seeking better performing workers and students, etc. The right to 'not enhance' should be firmly established, though it is obvious that it will be difficult to set limits on the creeping influence of marketing and advertising power and social pressure.

Reductionism and personal identity

All workshop attendees acknowledged that there can be no cognitive function without neural activity. In this sense the mind is the brain. But at an epistemic level - that is, as a knowledge system for understanding ourselves - 'mind' and 'brain' open up very different practices and experiences. This difference is a challenge for scientists working in different disciplines. And it might be of concern for European citizens too, since there is currently a tendency to *reduce* the mind to the brain. If the mind really is the brain, and the brain basically is just a bunch of molecules, where does this leave us? What is left of our personal identity? There is one consolation: this reductionism is as yet not based on facts. In fact, psychological insights are necessary even within a neuroscience framework, because we cannot understand the workings of the brain without 'mental' concepts. One should keep this in mind, especially now that 'brain talk' is becoming such a dominant cultural narrative.

Responsibility and Free Will

When working within the framework of the natural sciences, as many brain scientists do, one starts from the presumption that all events are caused by other, earlier events. This presents problems for our concepts of responsibility and free will - concepts that play an enormously important role in our social life, since they enable us to hold each other accountable for our deeds. But in what sense can a decision be 'free' if it is caused by former events? How can we be held responsible for deeds we could not help doing since they were caused by events which might have taken place long before we were born?

It is clear that the dominant way of thinking within brain sciences is at odds with our moral thinking, which relies heavily on the concepts of free will and responsibility. The findings of brain sciences might thus threaten moral and spiritual concepts which citizens hold dearly. The Amsterdam workshop made it clear that scientists and social thinkers, for their part, are trying to revise the concepts of free will and responsibility in ways which make them commensurable with scientific findings - and thereby save them for moral society. Scientists should acknowledge the theoretical impact of the fact that the brain is fundamentally a *social* organ, which moreover changes shape during the whole lifetime of a human being.

Privacy and personal freedom

The breakthroughs in the field of brain imaging are spectacular; it is becoming steadily less of a metaphor to say it is possible to look into the brain. This has obvious curative advantages. But the effects will be wider. Various sorts of screenings could be offered to the population, sometimes on market related grounds. This is worrying for privacy reasons; who will have access to this information, and with what purpose? The fact that some predisposition might in the future be detected by brain imaging techniques, such as one for Alzheimer's or schizophrenia, will impact an individual's prospects in life - and it might have repercussions for his ability to take out insurance. Screening techniques are already being explored for military and commercial use, notably in the United States. Neuro-marketing and 'brain-fingerprinting' techniques might invade the privacy of citizens and their right to keep their thoughts and feelings to themselves.

Ethics of research and information exchange

Citizens should be able to trust the fact that normal ethics of research are honoured in brain sciences, like the practice of informed consent and the precautionary principle. Scientists attending the workshop also acknowledged that the rapid progress of brain sciences presents a major challenge to them: their obligation to inform the public properly and promptly. This is not easy in a media-dominated world which is often at odds with the scientific method and the strictly rational mentality. Nevertheless, scientists-as-doctors should leave behind any remaining traces of paternalism, and scientists-at-large should make an effort to communicate with citizens.

All in all, the workshop made it clear that the topic of brain sciences has enormous potential for citizen deliberation. Its richness and interest as topic for discussion lies in its integrated, multidisciplinary approach - and in its many fascinating social, psychological and philosophical implications.

Part 1

| Workshop proceedings |

List of contributors

Professor Demosthenes Agrafiotis . Social Scientist . Hellenic National School of Public Health . Greece

Professor Jan Willem Berkelbach van der Sprenkel . Neurosurgeon . Utrecht University Medical Centre . The Netherlands

Dr James Butcher . Senior Editor . Public Library of Science . United Kingdom

Dr Axel Cleeremans . Director of the Cognitive Science Research Unit at the Free University of Brussels (ULB) . Belgium

Professor Johan den Boer . Biological Psychiatrist . University of Groningen . The Netherlands

Dr Berit Faber . Specialist in biolaw . Executive Director of the Danish Council of Ethics . Denmark

Mr Alistair Newton . Treasurer of the European Federation of Neurological Associations and Executive Director of the European Dystonia Federation . Luxembourg

Baroness Susan E. Greenfield . Professor of Pharmacology - University of Oxford . Director of The Royal Institution of Great Britain . United Kingdom

Professor Roland Jouvent . Médecin psychiatre . Institut Fédératif de Recherche des Neurosciences . France

Professor Flavio Keller . Developmental Neuroscience . University "Campus Bio-Medico" . Rome . Italy

Professor Detlef Linke . Philosopher and Clinical Neurophysiologist . University of Bonn . Germany

Professor Peter Molnar . Professor of Behavioural Sciences . Head of the Behavioural Sciences Department . University of Debrecen . Hungary

Professor Richard Morris . Professor of Neuroscience . University of Edinburgh . School of Biomedical & Clinical Laboratory . Division of Neuroscience . United Kingdom

Dr Paolo Lucio Morselli . Psychiatrist . Secretary General of the Global Alliance of Mental Illness Advocacy Networks . GAMIAN Europe

Dr Zoltan Nemes . Psychiatrist and Neuroscientist . University of Debrecen . Hungary

Professor Jochen Oehler . Neurobiologist . University Hospital . TU Dresden . Germany

Professor Jes Olesen . President of the European Brain Council (EBC) . Glostrup University Hospital . Denmark

Dr Ian Ragan . Executive Director of Neuroscience Research . Eli Lilly . United Kingdom

Dr Bernard Reber . Philosopher . 'Centre de Recherche Sens, Ethique et Société' . France

Dr Andreas Roepstorff . Medical Anthropologist . Aarhus University . Denmark

Professor Steven Rose . Neurobiologist . Director . Brain and Behaviour Research Group - The Open University . United Kingdom

Dr Wrye Sententia . Co-director . Center for Cognitive Liberty & Ethics . US

Professor Paolo Valerio . Professor of Clinic Psychology of the University of Naples "Federico II" . Director of the School of Specialization in Clinic Psychology . Italy

Professor Christine van Broeckhoven . Scientific Director . Department of Molecular Genetics . University of Antwerp . Belgium

Professor Mark A. van Buchem . Head of Neuroradiology . Department of Radiology . Leiden University Medical Centre . The Netherlands

| Aims and expectations |

The objective of the European Citizens' Deliberation initiative (ECD) on brain sciences is to design and implement a new way of interactive governance at European transnational level. Citizens from nine European countries will asses and deliberate on the issue of brain sciences with relevant research, policy and ethics experts, various stakeholders as well as representatives of European decision- making organisations. As such, the initiative aims to make relevant contributions to European policy-making and wider public debate.

Overture

But before involving citizens in the discussions on brain sciences, the members of the consortium that facilitates the ECD, felt the need to have an overview of the technological and social aspects of the subject. Therefore, two members of the consortium, the King Baudouin Foundation and the Rathenau Institute organised the expert stakeholder workshop, 'Connecting Brains and Society. The present and future of brain science: what is possible, what is desirable?'. The aim of the workshop was to investigate the technological state of affairs in brain sciences, the social consequences and the way in which they interact. Among the ECD representatives, the expectations for this overture were high. As Alessandra Drioli from the Città della Scienza in Italy put it: "I am sure this meeting will be very stimulating for our project and it will be exciting to find out about all the new things that happen in brain sciences." Graham Farmelo from the Science Museum in the UK even challenged the invited experts and stakeholders by stating: "Every one knows about the issues related to cloning, nuclear power and so on. What we need to find out in these two days from the experts, is why brain sciences should be an equally compelling issue so that the public can get involved in the subject."

The workshop was held on 22 and 23 April 2004 in the Shaffy theatre of the prestigious Felix Meritis building in Amsterdam. This location was not selected by coincidence. The building, located at the Keizersgracht, one of the famous canals in the centre of Amsterdam, was constructed in 1778 for the Felix Meritis society. The building and the society are expressions of the commitment of the citizens of Amsterdam to the advancement of the arts and sciences. Since the end of the 18th century, Felix Meritis has become a remarkable place for independent thinking, the transfer of knowledge and for enjoyment of the arts. It is a place where opinions were formed and where the results of the exchange of ideas on art, culture, science and society have contributed to cultural and political co-operation. At Felix Meritis, the practice of the arts and sciences and their impact on society, is still a matter for continuous exploration, whether this be through the processes of European integration or the development of international cultural networks (physical and digital). Could there have been a more appropriate location for the opening of the European Citizens' Deliberation initiative on brain sciences?

Two circles

For the workshop, a well-selected group of 25 European top-level scientists and stakeholders were invited to give their views on developments in the brain sciences from their own perspectives. Among them physicians, neurologists, psychiatrists, cognitive and social scientists, philosophers, artists and representatives of stakeholder organisations: the pharmaceutical industry, the European Brain Council, the European Federation of Neurological Associations (EFNA), the Global Alliance of Mental Illness Advocacy Networks (GAMIAN Europe), The European DANA Alliance for the Brain and the Federation of the European Neuroscience Societies (FENS) (see appendix 2 for 'Contributors Biographical Information').

The 25 experts addressed a group of 25 representatives from the ECD consortium. During the course of the workshop, the consortium representatives were often referred to as the 'outer circle', because they were sitting in a square formation surrounding the experts and stakeholders. The latter are therefore sometimes referred to as the 'inner circle'.

What and why

The term, 'brain sciences', does not have the same meaning for everyone. During this workshop and the whole course of the ECD project, the brain sciences are considered to be a field of study which deals with the structure, function, development, genetics, biochemistry, physiology, pharmacology and pathology of the nervous system. These pathologies include both neurological as well as psychiatric disorders. The study of behaviour and learning is also considered to be a division of brain sciences. The study of the human brain is seen as an interdisciplinary field which involves many levels of study: from the molecular level and the cellular level (individual neurons), through that of larger subsystems like the visual or auditory information processing level to large systems like the cerebral cortex or the cerebellum up to the highest level: the nervous system as a whole. But in the 21st century, the brain sciences also involve how we think about ourselves and how we deal with classical moral and societal concepts like responsibility and free will. Because brain sciences are not only about treating diseases, they form an important narrative about what it is to be human, as one of the participants in the workshop put it.

The field of brain sciences is rapidly gaining societal importance. There are several distinct but interlinking reasons for this. Firstly: the life-span of the European population has stretched enormously over these last decades. As individuals get older, demographic balances will shift, and many of the age-related neurodegenerative diseases like Parkinson's and Alzheimer's will increase dramatically. Combine this with the recent dramatic rise in reported psychiatric conditions and it becomes clear that there will be a huge demand for ways to alleviate or cure brain-related diseases.

There is also a scientific reason. There is a feeling within the scientific community that we are really beginning to understand the workings of the brain in a new and scientifically productive way. Neuroimaging, particularly functional techniques, facilitate insights into the living and working brain. The introduction of molecular biology techniques has started an equally important revolution in understanding the neurochemistry and the cellular functioning of the normal and pathologic brain.
And thirdly, the important theoretical advances in the cognitive neurosciences have provided the required level of analysis of these functions. All three fields have been boosted by the incredible increase in the power of computer sciences. This situation is completely unprecedented and it produces important spin-offs for diagnosis and neurosurgical or pharmacological treatment, and even for cyborg-type developments alike.

Bridging the river

The programme of the workshop was organised along the boundary between the technical and social aspects of the brain sciences. This may seem to be contradictory to the purpose of this workshop, but the organisers felt that this formal division was necessary to discuss adequately both the technological state of affairs and the expected social consequences. *One has to know both sides of the river before one can bridge it.*

On the first day, several experts in psychiatry, neuro- and psychopharmacology, genetics, imaging, surgery and cognitive neuroscience presented the latest and upcoming developments in their disciplines. All participants got the opportunity to ask questions and to clarify and discuss the content of the presentations. The presentations were a compromise between an exhaustive overview of the respective fields and strict timing. As a result, the speakers did not always have the time to expand on the technical details of some of the subjects and items which they mentioned. In those cases, the editors have provided a glossary as further clarification. An arrow in brackets [→ glossary] is used to refer readers to the glossary. The glossary is included in appendix I.

The second day of the workshop was devoted to the listing and mapping of the societal and ethical issues relating to the brain sciences. Twelve experts and stakeholders had been invited to prepare a short statement about what they considered to be the most pressing issues for citizens in Europe to debate. After each statement, the moderating science journalists recapitulated the issues with the speaker, with the help of the audience, grouped them into categories defined beforehand. At the end of this two-and-a-half hour mapping exercise, ten large notice boards (each representing a group of related issues), were completely covered with close to fifty different societal and ethical issues relating to brain sciences. After this mapping exercise, a voting session took place to decide which of the categories or issues would be further discussed in depth. The three topics that attracted the most votes ('limits of normalcy', 'reductionism', and 'responsibility and free will') were put to a debate immediately after the vote.

This workshop does not claim to be scientifically exhaustive, but is rather a dynamic exercise and a confrontation of opinions and feelings from experts with very different - and complementary - backgrounds. The listing of the issues, made collectively and extemporaneously, could naturally not be exhaustive either, and each item could have been further discussed for hours. The aim of the whole exercise was not to give a final answer to any question. It was really a meeting to raise questions, and to lay out the terrain for further exploration by European citizens. Or as Willy Weyns of the Flemish Institute of Science and Technology Assessment (Viwta) in Belgium put it: "With all these excellent brains present here, we have the core of a three-year project and we have found issues on which there is agreement between society, stakeholders and scientists, but also issues on which there is room for debate. There is definitely still a lot of work to do, but now we should be able to find headlines which we can propose to the public."

| Programme |

Moderators

Marjan Slob, Karin Rondia and Peter Raeymaekers, science journalists

Day 1: Thursday 22 April 2004

3.00 p.m. **Registration** at the Felix Meritis Centre

Opening session

4.00 p.m. **Welcome address**
Jan Staman, director of the Rathenau Institute
Introduction on the ECD initiative and workshop
Gerrit Rauws, director of the Governance Programme, King Baudouin Foundation
Short reflections of participants

Session 1

4.30 p.m. **State of the art as regards scientific, medical and technological developments in brain sciences and prospects for the near and distant future**
Format: Short presentations by experts in different technological fields and plenary discussion

Psychopharmacoloy and the future: beyond traditional boundaries
by The Baroness Susan Greenfield

Advances in psychiatry
by Professor Roland Jouvent

Genetic vulnerability to neurological and psychiatric brain disorders
by Professor Christine Van Broeckhoven

Cognitive neuroscience
by Professor Richard Morris

Brain imaging, the present, the future and societal consequences
by Professor Mark van Buchem

Neurosurgery
by Professor Jan Willem Berkelbach van der Sprenkel

Brain research deserves a better (funding) fate
by Professor Jes Olesen

7.45 p.m. Break

Session 2

8.15 p.m. **Plenary discussion concerning the possible role of brain science in society and future developments in brain science**
Moderated by Jan Staman

9.00 p.m. **Dinner**

Day 2: Friday 23 April 2004

Session 3

9.00 a.m. **Views on developments in brain science from a societal perspective**
Format: Short statements on societal issues by several experts and stakeholders followed by questions, plenary discussion and categorisation of the issues

Statements were made by
Professor Steven Rose
Dr Zoltan Nemes
Dr Berit Faber
Dr Ian Ragan
Professor Paolo Lucio Morselli
Mr Alistair Newton
Dr Wrye Sententia
Professor Flavio Keller
Professor Jochen Oehler
Professor Demosthenes Agrafiotis
Dr Andreas Roepstorff
Dr Bernard Reber

11.45 a.m. **Voting session**
Format: After purposeful ranking of the societal issues, the most crucial ones are selected for further discussion

Session 4

12.15 p.m. **Plenary discussion of the 'Limits of normalcy'**

1.00 p.m. Lunch

Session 4 continued

2.00 p.m. **Plenary discussion of 'Reductionism'**

2.30 p.m. **Plenary discussion of 'Responsibility and free will'**

Closing session

3.00 p.m. **Plenary evaluation of the outcomes of the first European workshop 'Connecting brains and society'**
Moderated by Jan Staman

3.30 p.m. **General conclusions and close**
Gerrit Rauws

4.00 p.m. **Informal closing of the workshop with a social drink**

Session 1

| State of the art as regards scientific, medical and technological developments in brain sciences and prospects for the near and distant future |

Introduction

The first session of this workshop was devoted to an overview of what brain science [→ glossary] is and what the scientific, medical and technological developments are in this field of science. Progress in brain science is not limited to just one discipline. It is nurtured by developments in a very broad range of scientific and technological fields like genetics, physiology, biochemistry, pharmacology, surgery and medical imaging. We therefore asked seven distinguished speakers to give a state of the art survey of their fields in the brain sciences.

This session will be opened by the Baroness Susan Greenfield who will talk on the future of neuro- and psychopharmacology. Secondly, Professor Roland Jouvent will give his view on advances in psychiatry and Professor Richard Morris will give an overview of the cognitive neurosciences. After the break, Professor Mark van Buchem will talk about the present, future and societal consequences of brain imaging, followed by Professor Jan Willem Berkelbach van der Sprenkel who will consider the evolution of neurosurgery. Session 1 will be closed by Professor Jes Olesen who wants to stress how brain research deserves a better fate.

Unfortunately, Professor Christine Van Broeckhoven, who was to talk about the role of genetics in neurological and psychiatric disorders, could not attend the workshop. Since we were only notified of her absence only a few hours before the start of the workshop, the organisers were not able to replace her in time. Professor Van Broeckhoven has sent in an extended abstract of her presentation for these proceedings. Furthermore, all other speakers were asked to include, if possible and applicable, their thoughts on the impact of the expanding knowledge of genetics on the field of the brain sciences.

Each speaker was asked to be succinct and brief and to keep their presentations within a time frame of 15 minutes. But at the same time, they should be exhaustive in their overview, predict what the developments in their field could bring us in the future and what the possible social consequences of these developments might be. Furthermore, the talks should not be intended for their fellow scientists but should be at a level and in a language comprehensible to lay people ... all of these preconditions must have truly been a daunting task for the speakers. As a result, the presentations are a compromise and sometimes the speakers did not have the time to expand on the technical details of some of the subjects which they mentioned. In those cases, the editors have provided 'information boxes' which present further clarification. However, it should be kept in mind that this information was not necessarily discussed during the workshop. For the sections for which these boxes are available, a footnote indicates the relevant box.

After each of the presentations, a plenary discussion allowed for further clarification and gave other experts the chance to give their views. These discussions focussed mainly on the scientific, medical and technological aspects and to a lesser extent on the social issues. The discussion of those issues was reserved for the second day of the workshop.

Psychopharmacology and the future: beyond traditional boundaries

BY THE BARONESS SUSAN GREENFIELD

The Baroness Susan Greenfield is Professor of Pharmacology at the University of Oxford and Director of the Royal Institution of Great Britain. Her research is aimed at developing strategies to prevent neuronal death in Alzheimer's and Parkinson's disease. The Baroness is also a media personality with numerous appearances on television and contributions to newspapers. She has written several books for the lay public and she was nominated woman of the year 2000 by the Observer.

Pharmacologists are living in an exciting time because they are now going beyond the traditional boundaries of neurotransmission. Neurotransmission is the transfer of a signal from one neuronal cell to another using a chemical messenger molecule. This messenger is called a neurotransmitter. Neurotransmitters are needed because brain cells cannot transmit their electrical signals to other cells directly; they need some kind of a 'go-between'. This is comparable to when you drive a car from Great Britain to the Continent. You cannot cross the Channel with your car. To get across it, you have to transfer to another mode of transportation: a boat or a train. The synapse, the gap between two neurons, is the neuro-anatomical counterpart of the Channel. Upon arrival of an electrical signal on one side of the synapse, the neuron has to switch to another means of transporting the signal. This alternative is a chemical messenger in the neuron on the other side of the synapse. The transmitter interacts in a kind of handshake with the receptor, as a key in a lock. The molecular handshake will trigger a new electrical signal in the next cell.

Beyond transmission

Until now, neurotransmission is really the way in which the medical world has dealt with brain disorders. Drugs were developed with the aim of manipulating these transmitters either by increasing their availability, by blocking their action or by replacing them.

But there are complications with this approach. Firstly, some transmitters act like fountains in the brain. This means that one transmitter is not responsible for one brain function. So if you want to target, for example, the dopaminergic system in Parkinson's disease, you are targeting other systems in which that chemical is at play as well. Dopamine is one of the best known and one of the first discovered neurotransmitters. In patients with Parkinson's disease there is a shortage of dopamine in a specific region of the brain called the substantia nigra. A logical treatment is to enhance this shortage by providing more dopamine. However, one should be aware that at the same time too much dopamine in the brain is associated with schizophrenia. So, as long as one is targeting one neurotransmitter, one always will have side effects.

A second complication is the existence of capricious receptors. Actually, the metaphor of the key in the lock is not really a good metaphor for the interaction between neurotransmitter and receptor. A better metaphor is the shaking of hands. If you shake hands with someone and you squeeze the hand of your partner, after a while the hand becomes numb. That is also what happens to these receptors. They are not invariant in how they respond to a transmitter. What will happen is that they might desensitise, just like your hand, or on the contrary they might become very sensitive. So the interplay between neurotransmitter and receptor is changing all the time. This means that also the action of drugs, which are working on those receptors, changes as well. Therefore, if you treat a brain disorder over a chronic period of time with a drug, it is hard to predict a uniform response.

Another problem is that targeting the transmitter might lead to alleviating the symptoms, but not treating the primary cause. If we take the example of Parkinson's disease again, by giving dopamine back to the brain, one does not prevent the brain cells from dying. And it is cell death that is the primary cause of the disease. A better treatment would be looking at ways to stop those cells from dying, instead of just supplying the brain with dopamine. The supply of extra dopamine will only alleviate the symptoms and not treat the cause of Parkinson's disease.

However, all these arguments do not mean that further developments in the science of neurotransmission are no longer interesting. On the contrary. We are increasingly discovering new subtypes of receptors and we have also found out that receptors have multiple binding sites, so called allosteric sites. These sites modify the binding of the neurotransmitter on the receptor. A strategy, on which a number of pharmaceutical companies are embarking, is to get more specific chemicals which target one specific subtype of a receptor or which modify the interaction between transmitter and receptor through interaction on a very specific allosteric site. A very well-known example of this approach is benzodiazepine (Valium), which binds on an allosteric site from a particular receptor.

Molecular biology

But the most exciting things in neuropharmacology are happening beyond neurotransmission. So what is there beyond neurotransmission? First of all, there is the development of molecular biology [→ glossary]. Thanks to increased knowledge of the human genome, molecular biologists promise us that we are evolving from about 400 molecular targets to developing drugs against over 4 000.

Secondly, molecular biology is also providing us with a new class of drugs, besides small chemical molecules. These new medicines are proteins. Unfortunately proteins don't get easy access to the brain because of the blood-brain barrier which makes it unclear whether protein drugs will be useful in neuro- and psychopharmacology. Nevertheless some people believe that proteins like growth factors and trophic factors - which enable brain cells to grow - might be useful in cases of brain cell death. Neuromodulators have also been very popular for the last ten or fifteen years. They don't necessarily have a direct action on neurotransmission, but they put the cell on red alert. They sensitise the cell, they make it somewhat easier when the signal comes along.

A less glamorous, but particularly useful development are surrogate markers. In terms of neuropharmacology, this development is underestimated. Surrogate markers allow monitoring the course of a disease in an easily accessible body fluid like urine or blood. With these devices one can mark out a person's individual course of a disease and predict how the disease will evolve: how long it will take before the patient needs a wheelchair, when a person will need to be institutionalised etc. Even though there might not be a treatment or a cure for that disease, these predictions allow people to plan out their lives. Also for the pharmaceutical companies these surrogate markers are very useful. They would increase the speed of drug discovery and would make drug development much cheaper also.

New hypotheses

Most non-neurotransmitter targets originate from alternative and novel hypotheses in brain science. One of those hypotheses, and my personal favourite, is that neurodegeneration is an aberrant form of development. In the developing brain there are lots of neurotrophic factors [→ glossary] that will enable cells to form certain configurations. But once you become an adult, these factors are not so active anymore in most populations of neurons. So in an adult brain, so called serial cells, which form neuronal circuits, gradually loose their plasticity and capacities to proliferate. In the case of brain damage, these serial cells get compensated by extant cells, which won't grow again. On the other hand the population of global brain cells, so called because they act like fountains, retain their plasticity, sensitivity to neurotrophic factors and regenerative capacity. It is possible that this potentially beneficial capacity becomes toxic when it is wrongly activated and precisely this might go wrong after damage to an adult brain. The landscape of the adult brain might have changed so much that only erroneous activation takes place in the case of brain damage. I discussed the last example merely to indicate to you one of many alternative hypotheses, which could lead to new approaches in brain disease beyond the increase or decrease of neurotransmitters.

Beyond genes

If current neuropharmacology is to look beyond transmission, I make a case that we should also look beyond genes. Despite the hype of the

1990's, it is really a bit of a far cry to jump from DNA to the structure and function of the brain. If scientists talk about the gene for this and the gene for that outside of labs and especially in the media, it is precisely this jump they are making. It is crazy to think that 'good housekeeping' or 'being witty' is trapped in a strand of DNA. At this moment, nobody really knows how genes relate to brain function. All that genes do, is make proteins and proteins contribute all over the brain in many different ways, contexts and situations.

If we talk about genes, we immediate think of gene therapy [→ glossary]. But the problems with gene therapy are immense. Even if you know in which gene the mutation is located, the targeting of that gene is problematic. When for example the gene for cystic fibrosis was identified more than ten years ago, everybody was really excited that the mutated gene in CF-patients could be corrected by gene therapy. It turned out that targeting these mutated genes is far from easy, especially since in humans we are legally not allowed to change genes in egg cells and sperm cells. Therefore one can only transfect somatic cells in the body with correcting genes. Some researchers are now able to do that with viral vectors. Others use bolistics, which is DNA mixed with tungsten particles which are 'shot' in the cells by high powered guns.

But even if we technically would be able to convey correcting genes to diseased tissues, it remains unclear what are we going to achieve for brain disorders when we change a gene. I think there is an over-interpretation, an over-hype, of the correlation between genotype and phenotype. A classical example is that of Huntington's disease. The disorder is caused by a mutation in a single gene. But even in this relatively 'simple' genetic disorder, environment plays a key role in the onset and the course of the disorder. Research has pointed out that the genetic destiny of transgenic Huntington mice, which all carry the mutation which causes Huntington disease, is changed by simple manipulation of their environment. By enriching their housing conditions with 'play tools' the degree of onset is much more modest and the age of onset is much later compared to a standard housing. So the genotype-phenotype correlation is much more complex than just one gene, one behaviour. Even in a mouse. Even in a pure single gene disorder like Huntington's.

For Alzheimer's disease, it would even be more complex since there are different genes which have been associated one way or another with the disorder. Unlike for Huntington's disease, there is no 'single gene for Alzheimer's disease'. You do not have Alzheimer's disease trapped in a specific gene. So we have to be very careful if we extrapolate from genes to function in the brain.

Another problem with genetics is the problem of model organisms [→ glossary]. People are very excited about animal and genetic models (transgenic and knockout models). Again, if we take the example of

Alzheimer's disease, there are hundreds of different mouse models where people think they have found the Holy Grail for Alzheimer's disease. But a model implies that you know the salient feature of the disorder and that you can extrapolate it from all secondary effects. For Alzheimer's disease we do not know what the core problem is. According to some, it might be a protein called amyloid, but I personally do not believe that. It is at least highly controversial. So if we do not know the core of the problem, how can we model the problem?

I hope I made it clear that, if we are going beyond traditional boundaries, we should be going beyond transmitters, beyond genes and also beyond the molecular level because the brain is more then the sum of its molecular parts. If you just say the brain is transmission, genes or proteins, than you miss out on the most important issue about the brain, namely the connectivity of the neurons. This connectivity represents the higher order structure of the brain. If you were to count, at a pace of one per second, all the connections in the brain with its 100 billion neurons, each with roughly 100 000 connections, it would take you 30 million years. Furthermore this connectivity is a truly dynamical process. The brain is a plastic organ which changes all the time. This continuous change, this brain plasticity [→ glossary] is precisely the reason why we can learn new things and reinforce the remembering of things of the past.

Stem cells

Another new approach for treating brain disorders is stem cells [→ glossary]. These are cells in our body, which can become heart, muscle, lung, blood and brain cells. But I want to caution against the hype which is currently being created about stem cells. In theory one could indeed treat Parkinson's disease with dopamine cells. But the neurosurgeon would have to deliver these cells in the substantia nigra, which is a nucleus located very deeply in the brain. So access of the target tissue is the first problem one has to deal with. Secondly, there is also the problem of tumour formation. Stem cells are dividing cells, and an uncontrolled proliferation of those cells might be problematic. A third problem is the regulation of the dopamine levels produced by these cells. With an overproduction, these patients might be at risk from schizophrenia.

Chips in the brain

There is also the interfacing of silicon with carbon that in ways can intercept brain organisation. Human nerves can grow on silicon circuits and form connections. Neurons generate these small electrical currents, which make them brilliant electronic components. Neurochips have a future, no doubt about that. Some people think we might send and receive our e-mail directly from and to our brain, which will be connected to a chip. As an alternative to typing it on a keyboard and reading it from a screen.

Personally, I do not think that that will be possible, although there is definitely a future for direct connections between the brain and the computer. Chip implants will be very helpful for people who are quadriplegic. These developments trigger, of course, ethical and philosophical issues about free will, mind and brain ... and all those other traditional dichotomies.

Nanotechnology

The ultimate interface between carbon and silicon is nanotechnology [→ glossary]. Nanoscience in the brain has very practical and current technology applications like DNA probes with which you can see whether certain genes are active at certain moments, diagnostic antibodies that can bind to magnetic nanoparticles, etc. Nanotechnology can be a whole new revolution.

Beyond disease

Briefly then to a future which goes beyond brain disease. We all hope to live longer and healthier lives, increase leisure and expect more happiness. Don't we all think now that we should be happy? In the previous generation, people just wanted to bring up their kids, have enough to eat and have a warm house to live in. Now expectations have changed enormously ... but with consequences!

We talk about having a sensational time. A lot of people take this literally. For them this means a world where there is a premium on the senses. It is a world stripped of all cognitive content. It is a world without meaning. There is just a flood of flashing lights, pounding music, a beat of techno, a premium on smell and touch ... back into the booming, buzzing confusion of the infant. And, very intriguingly, we pay money to do this. Whether it is skiing, dancing, drinking or perhaps more refined pursuits in the search for sensations, what they all have in common is that we are the passive recipient of our senses and we have abrogated the sense of self. In a mind state like that, there is no longer self-consciousness, you have let yourself go, you are out of your mind, you have lost your mind, you have blown your mind ... the very word ecstasy in Greek means 'stand outside of yourself'. It is a very interesting aspect of the human condition that we oscillate between wanting to become individuals and having high status, but by the same token, from time to time doing this.

Since we have this desire to be so happy now, and it seems that it is a premium to be in the booming buzzing confusion, there are a number of drugs on the market, which cross the boundary between health and lifestyle. And this is a very important issue: where do we start saying something is an illness and where do we say it is lifestyle. When are we drifting towards the 1984 world of George Orwell, in which people go back to being in the here-and-now and somehow have lost some aspect of their humanity?

Bodies and brain growing older

Every one of us evolves from baby, to adult, to senior. The marvellous thing about getting older is that your brain, if it is properly stimulated, will flourish and become more and more you. Increasingly, as we live in the information age with less emphasis on physical strength, older people will become more important in our society. They will be working more and they will be more active than ever before. In that society, the healthy brain and what it can do will be very much at the fore of people's discussion.

Sunset or sunrise?

We are in a sunset or sunrise scenario. In the sunset society there will be an overemphasis on genetic determinism. People want to be better than other people and take enhancing drugs and perhaps enhancing gene therapy. These will make them better but not happier. In the sunrise society we use brain science and technology to really understand the exciting questions of what makes us individual. But above all, we use our knowledge to proselytize the notion of individuality across the society as a whole.

Discussion

IAN RAGAN: You made some very good points about the limitations of pharmacology. I agree that the traditional small-molecule approach in pharmacology has certainly limitations. Indeed in a number of cases, as in degenerative diseases, they rather treat the symptoms instead of the cause of the disease. To get over the limitations of the small-molecule approach what we are looking for in the industry is to extend the number of approaches to treating diseases and that includes gene therapy, anti-sense and DNA-approaches and many other things.

But another interesting point is that I think you overestimated the number of possible drug targets. With the human genome now at hand, we realise that the human being has far fewer genes than we once thought. We imagined that humans had far more genes than any other organism. That was snobbery on our part because the human genome turns out to have only about 30 000 genes. You pointed out that about 400 to 500 of them are now targets for drug discovery. Estimates have been made that the number of genes, which would be targetable for drugs, is at most 3 000. Those are the ones against which we could design molecules. Secondly, you have to guess which genes might be involved in disease processes. That might be about 3 000 as well. The big question is to what extent those two are overlapping and we don't know that, but it isn't going to be a lot. Anyhow, much fewer than 4 000.

SUSAN GREENFIELD: Sorry if I gave the wrong impression. The 4 000 that I talked about are not just gene targets. There are other types of targets as well, like parts of proteins, small peptides, which have their own actions.

Advances in psychiatry

BY PROFESSOR ROLAND JOUVENT

Professor Roland Jouvent is a psychiatrist who teaches at the Faculté de Médecine de la Pitié-Salpêtrière in Paris, France. He also leads a National Research Unit which studies vulnerability and adaptive behaviours. His group is involved both in animal experiments and human studies. His interests are the psychopathology, psychobiology and psychophysiology of depression and the action of antidepressants. He is also interested in the interaction of new technologies, like virtual reality, with psychiatric pathologies and/or society.

As far as the future of psychiatry is concerned, my opinion is that the problem is not finding new compounds (with the possible exception of gene therapy), but introducing a new global and logical approach that could combine psychopharmacology and social therapy.

The actual psychopharmaceutical agents, a theme which Susan Greenfield has brilliantly developed, have a real efficacy as regards symptoms, but they are not very successful in the social adjustment of patients, which is the main challenge for psychiatry which I see in the near future.

Other problems concerning drugs are the existence of large discrepancies as regards their efficacy, and inability to cope with social dysfunctions like unemployment, violence, aggressive behaviour or sexual abuse. The answer to all of these problems is probably not to be found in some new compounds but rather in a better understanding of this large part of the psychiatric field which concerns social interactions. So my presentation will focus on this single point, and I hope you'll find it of interest.

Imagination and action

Twenty years ago, we were all working on the biological understanding of the mind, and now progressively, after a purely cognitive state, we are entering a new era by trying to understand social cognition and social behaviour from the cognitive and emotional point of view. So I think this is very important, because if we are able to understand how we deal with others, it will affect everyone, and not only the people with mental disease.

I think many social misunderstandings are rooted in people's failure to recognise how much their perceptions of situations may differ from that of others. This misunderstanding is a main point of the psychiatric consequence of disease. And I will show you some recent data that could help link social dysfunction to the neuronal one.

We all know that some cerebral activation occurs when you lift your arm. Surprisingly. the same cerebral activation occurs when you imagine that you lift your arm. You can see that the same neurons are activated by an action and by the imagination of an action.

This is very economic from the Darwinian point of view; it would have been stupid to have two sets of neurons, one for the action and one for imagining it. Isn't it fascinating to realise that thinking is doing? This might be a special ability of humans and perhaps also of primates.

Mirror neurons

Moreover, when you see another person doing an action, again, you activate your own neurons implied in the execution of this action. This is the theory of 'mirror-neurons': similar brain areas are activated while representing one's own action and while observing the same action performed by another person. When you look at somebody else, you simulate the same action to understand it. I think this is a very important contribution to making the bridge from the biological cognitive brain to the social one.

So if you accept that social cognition is based partially on this simulation phenomenon, we can agree to target the dynamics of interpersonal behaviour for new research in the future.

These mirror neurons are also described in emotional exchanges. When you ask subjects to inhale odours producing disgust, you have the same neuronal activation as when people watch video clips showing a facial expression of disgust. So once more, in order to understand an emotion occurring in others, we have to simulate it in our brain. We simulate the emotion to understand the emotion of the other person.

So, a general feature of human cognition is to see others as similar to ourselves. And one important point concerns the distinction in the simulation activity between me and others. If I simulate that you are me, I have to get information saying that "me is me", and not you. Some parts of the brain, particularly the inferior parietal right cortex, have been showed by Jean Decety (from The Institute for Learning and Brain Sciences at the University of Washington in Seattle) to be activated during the process of simulating actions and to remind the person of their individuality. We know that schizophrenic patients, as well as autistic children, lack this faculty. We call this faculty 'agentivity', but we don't know the real mechanisms leading to it. Perhaps it also functions in the psychotherapeutic process, but this is still an unexplored topic.

Drugs versus psychotherapy

Classically, in psychiatry, there is this so-called opposition between pharmacotherapy and psychotherapy [→ glossary]. The primary target of pharmacotherapy is the ancestral part of the brain (limbic system, basal

ganglia, substantia nigra, etc.) whereas psychotherapy acts mainly upon the associative networks, in the mature part of our brain. In my opinion, there is no antagonism between the two approaches, and a combination of both forms of treatment is probably the way for the future.
In the psychotherapeutic process we can distinguish several levels of action. The cognitive approaches are very sophisticated, playing on the representational level. The psychoanalytic ones might rely on more primitive levels because they combine the Freudian impulse (the biological tension) and representation. And now with new therapeutic approaches like EMDR (eye movement [→ glossary] desensitisation and reprocessing) or virtual reality as a therapeutic tool, we have new types of psychotherapy which combine mental, representational and physical subcortical targets.

But whatever the treatment, whatever the classification, it is fundamental to understand that there is always a reciprocal simulation of patient and therapist during the exchange, because to understand each other they have to simulate each other. It's a difficult part for the therapist to distinguish between his logical, medical part of the brain, and his compassionate one. So the new challenge is probably to incorporate in the education of students the simulation of the other person as a concrete part of the treatment, whatever the treatment.

Virtual reality

Finally, for two reasons, I'd like to make one point about virtual reality. Virtual reality is becoming a treatment for many disorders today (agoraphobia, claustrophobia, etc.) and probably in the future for more psychiatric diseases. The capacity to be immersed under the control of a therapist, who can see what the patient sees, is a good alternative to cognitive therapy. But the second reason why I want to mention virtual reality is because of the risks involved in it. If you compare a depersonalisation scale and a presence scale (which is used to measure the presence of normal subjects in virtual reality), you can see many similarities between practice of the phenomenology of virtual reality and the beginning of depersonalisation. So the subjects at risk of schizophrenia are probably particularly vulnerable to the practice of this new successful technology.

I would like to finish by showing you this photograph of the Pont des Arts, from Robert Doisneau, to show you that I think the psychiatrist or researcher is in a position to look at the picture the patient paints of the world. And so he has to perform the double action of simulating: the real world with what the patient sees, or thinks, but also his own simulation of the world.

Discussion

STEVEN ROSE: I want to make a point about one of your last slides where you made a link going from gene to communication. You argued that we need to ensure that there is a unity of discourse between social psychology and genetics. I think that one has to be very careful not to assume, as Wilson does, that the primary explanation lies at the genetic level. Consilience [→ glossary] to Wilson means subservience of all the other levels to that level. I think we have to be very careful in this meeting not to assume that what you talked about as high levels are subservient to lower levels. Consilience works both ways.

ROLAND JOUVENT: Yes. Perhaps we can assume that probably genetics is at the end of its first phase of what concerns psychiatry and that reductionism is finished. Nobody thinks that psychiatric diseases are Mendelian or monogenetic anymore. We now see the genes as vulnerability factors, not as regards a disease but as regards one dimension of the evolution of a disease, for example, negative symptoms or aggressive symptoms. In alcoholism, some genes are known not to predispose to alcoholism as a whole, but to a special type of brain evolution in alcoholism. And so perhaps we can keep one part of the genetic model, not as an etiopathogenic model but as an explanation of the way in which the disease evolves.

STEVEN ROSE: What we need in this area is to transcend this nature/nurture dichotomy which belongs to the 19th century. And to do that we need new theories, and what is called developmental systems theory - which actually goes beyond nature/nurture - is trying to partition out in that sort of way. But there is another problem as well with what you list there. Because that's what we might aspire to, but I think we have a fundamental problem in the brain sciences at the moment, and it is that we cannot go between one level and the other. The language of molecular biology and the language of cognitive psychology, and the language that you were speaking, are much further apart than French, English, Danish or Dutch!

WRYE SENTENTIA: I'm curious about the distinction in the law, where action and thought are sparking at the same places in the brain. Because in the legal system at least in the US, the defining line of interaction on a legal basis is when you act. We already have, unfortunately, cases of pre-emptive war going on. I'm curious about pre-emptive restraint on thought.

ROLAND JOUVENT: Well, this is a big question. Would you think that simulation is a way to better reliability of results in psychiatric practice, or not? Isn't it a way to a better definition of action? As a therapist, my opinion is that we now dispose of enough scientific results to introduce psychotherapeutic practices in the field of science. Probably, it is not yet complete, but it's important, particularly in Europe and in France, where some types of psychotherapy are very 'magic' or 'mystic', for the evaluation and the formalisation of teaching. There is major confusion surrounding psychotherapy. If we take one part of scientific reasons as a transitory model, perhaps we can have an agglomeration of various types of schools of psychotherapy which could join the other parts of psychiatric practitioners and researchers.

KARIN RONDIA: Would it be possible to assess the results of psychotherapies?

ROLAND JOUVENT: Traditionally, there are scales but these are the same scales as for the results of psychopharmacology. So it's not a good way. It's very difficult to assess the new feelings of a patient as criteria for good treatment. But it's possible now to describe some models of formalisation of the action of the therapist.

STEVEN ROSE: We've actually done that. We've measured the effects on people undergoing non-drug based psychotherapy, on a scale which is normally used for assessing the effects of antidepressants. You can show that, as people undergo psychotherapy and their depression lifts, the biochemical measures lift as well. It's a very nice demonstration of the interaction between the therapy and biochemistry.

ROLAND JOUVENT: But that's for behavioural therapy, not for mind therapy.

PAOLO VALERIO: I think we are discussing some very difficult issues. If my mother dies, am I depressed? I am sad. Do I need drugs or do I need religion, or do I need psychotherapy? This is a big issue among us. Now, very often, because of medicalisation, mourning is not possible anymore. Look, I wear black. In my culture, black was used by women all their life for mourning their mother, father, husband, etc. But now it is just being elegant!

ROLAND JOUVENT: If you are mourning, you are sad, you feel blue. But if you have insomnia, lack of energy, inability to concentrate, then you need some drugs. They will give you a great benefit in terms of delay of action. So I don't think there is competition between the two. If you take only drugs, you won't work out your mourning within your mind. You might be relieved of vegetative symptoms but you'll still have the mental ones. With drugs, in this case, you won't get relief from your grief, even if you have more energy. Do you agree? But we can discuss this further.

BERIT FABER: I would like you to elaborate a little bit more on the use of virtual reality in combination with cognitive therapy. I find this very interesting.

ROLAND JOUVENT: My opinion is that within five years, not only virtual reality but also robotics as a support for child rehabilitation or for cognitive rehabilitation on computers will be 30% of treatment in psychiatry, because it is very efficient and economical. Virtual reality is interesting because we can make the hypothesis that in the evolution of a world with more and more virtual pleasures and activities, pathologies will change. There is plasticity in the phenomenology of mental disorders, and I think we will probably see new expressions of pre-schizophrenic symptoms or of delusions in adolescents in the coming years. So I think it is important to be able to understand how the virtual world is soliciting the self. When I'm sitting on my chair and I'm simulating that I drive a very fast car, do I have to leave behind the idea that 'me is me' to get into the virtual world? Do I have to leave my personality, my identity behind? That's a challenge for the future of our children. These evolving pathologies should be anticipated by new types of investigations, and probably virtual reality will show itself to be superior to classic psychotherapy for subjects with poor mental imagery. If you are suffering from claustrophobia, I will tell you to imagine a lift. But if you can't imagine the lift, the therapy will be ineffective! For those patients who can't make a mental image of their subject of fear, you can immerse them progressively in a world in which you can control the targets which change of their feelings and their beliefs.

Genetic vulnerability to neurological and psychiatric brain disorders

BY PROFESSOR CHRISTINE VAN BROECKHOVEN[1]

Professor Christine Van Broeckhoven is a molecular geneticist. She is scientific director of the Department of Molecular Genetics of the Flanders Institute of Biotechnology, Professor at the University of Antwerp in Belgium and Research Director at the Born Bunge Foundation.

Professor Van Broeckhoven's team works on identifying the genetic factors involved in various neurological diseases such as Alzheimer's dementia, bipolar disorder, Parkinson's disease and idiopathic epilepsies. She and other members of her team have been awarded several scientific prizes.

With age, vulnerability to brain disease increases significantly. Since in industrialised countries we are dealing with an ageing population, brain diseases are becoming very rapidly a major socio-economical burden.

Nature and nurture

The traditional methods of studying the involvement of genetics in brain disorders are family, twin and adoption studies. On the one hand, family studies make it possible to see whether familial clustering of a disorder occurs, while twin and adoption studies make it possible to tease apart the effects of shared genes, a shared environment and an individual environment. From these types of studies, it can be concluded that the majority of the neurological and psychiatric brain diseases are multifactorial in nature, implying both genetic and environmental factors underlying the risk of disease.

Genetic factors involved in dementia

For neurodegenerative brain diseases such as Alzheimer's dementia or Parkinson's disease, molecular geneticists have successfully identified several causal genes using a positional cloning strategy in multiplex families. In these rare families, the disease segregates according to a clear Mendelian inheritance pattern and results from a genetic defect in a single gene.

In the case of Alzheimer's dementia (AD), most patients belonging to these multiplex families develop the disease before 65 years of age and are therefore called early-onset AD cases (EOAD). To date, 3 genes responsible for EOAD have been identified: the amyloid precursor protein gene, presenilin 1 and presenilin 2. Presenilin 1 is the most frequently mutated gene with a mutation frequency of 18% to 50% in autosomal-dominant EOAD.

The identification of genetic risk factors for AD patients who do not belong to these multiplex families has been less productive. One major risk factor associated with neurodegenerative dementias is the apolipoprotein E isoform E4. Many studies reported other susceptibility genes, but the APOE epsilon4 allele has been the only risk factor that was consistently replicated in all AD populations.

Extensive cell biology research over the past ten years has led to the hypothesis that the four genes mentioned lead to AD through a common biological pathway resulting in abnormal processing of the amyloid precursor protein by subtly different mechanisms. Now, transgenic mice are being produced to study the influence of mutations in these genes *in vivo*, which will possibly lead to the development of novel therapeutic strategies.

Parkinson's

Several genetic factors involved in autosomal-dominant forms of Parkinson's disease have been identified. These include mutation in or duplication/triplication of the alpha-synuclein gene and mutations in the UCHL1 gene. Also several loci for autosomal-recessive early-onset Parkinson's disease have been identified, including mutations in the parkin gene and in the DJ1 gene. There is also evidence that mitochondrial mutations may cause or contribute to Parkinson's disease. A mutation in the gene-encoding synphilin-1 was found in 2 patients with sporadic Parkinson's disease.

Psychiatric diseases

Genetic studies of psychiatric diseases have moved ahead much more slowly but recently major breakthroughs have also been achieved, particularly in schizophrenia where currently three genes are under study. Genetic analyses of bipolar disorders, on the other hand, have not yielded any major risk gene at the moment. It is clear that genetic studies of these complex brain diseases suffer from many problems inherent in both the phenotype (e.g. phenotypic and clinical classifications) and the genetic tools (e.g. genetic and statistical methods of analysis of large data sets) currently available.

Genetic profiling

However, one can expect that, with a better understanding of the functioning of the human genes and genome, these complexities will also be overcome and that genetic profiling of major brain diseases will become a reality.

Knowing an individual's genetic risk profile might be helpful in designing more efficient prevention and/or treatment strategies. Genetic profiling is therefore one of the current major topics in the pharmaceutical industry (e.g. pharmacogenetics [→ glossary] and genomics) for the purpose of testing drug efficacy and potential negative side effects against the person's genetic background.

---

1 Professor Van Broeckhoven was unable to attend the workshop but she sent in an abstract of her presentation.

Cognitive neuroscience

BY PROFESSOR RICHARD MORRIS

Richard Morris is a cognitive neuroscientist who studied natural sciences in Cambridge. After holding several positions in Sussex, Durham and at the Massachusetts Institute of Technology, he is now head of the Laboratory for Cognitive Neuroscience at the University of Edinburgh. His ambitions are to forge closer links between the basic and clinical neuroscience communities.

He is also teaching at the Norwegian University for Science and Technology, and serves as the Life Sciences Coordinator of the UK Foresight Cognitive Systems Project. He is a Member of the Council of the European Dana Alliance for the Brain.

The focus of his research is on the function and neural mechanisms of the hippocampus and its role in memory. His research team tries to find out which neural mechanisms are engaged to encode, store, consolidate and retrieve information.

First of all I would like to congratulate Jan, Gerrit, the KBF and the Rathenau Institute for having the imagination to organise a symposium like this. It comes at a timely moment. There have been sufficient advances in brain science so that scientists actually have something to offer in this process of dialogue with the public. I hope that today and tomorrow we will be able to bring out some themes that the members of the ECD-consortium can take home with them.

My task today is to talk about the science of cognition. Cognitive neuroscience represents a major subcomponent of neuroscience. It focuses on the neural basis of information processing by the brain. This definition is not very limitative. It does not leave out a lot. But I think that this definition describes very well how cognitive neuroscientists see themselves. If we look at textbooks about cognitive neuroscience, for example at Gazzaniga's Second Edition of "The New Cognitive Neurosciences" (MIT Press, 2000), then one sees chapter headings like 'Development, Plasticity, Sensory Systems, Motor Systems, Attention, Memory, Language, Higher Cognitive Functions, Emotion, Evolution, Consciousness'. These headings give a flavour of what we are talking about.

From a new astronomy ...

What could be the basis of public interest in the brain? From the perspective of a cognitive neuroscientist, let me offer some suggestions - at various levels of complexity. I hope that during the course of this meeting, others will add substantially to these. At a simple level, perhaps from the point of view of a director of a museum through the corridors of which many children walk, one could think of the neurosciences as the new astronomy. Humans have long gazed at the stars and wondered about them; they were searching for the position of the earth in the solar system and beyond but at the same time they were looking to find their own identity. Now a great number ponder the intricacies of mind and brain in the same way.

... to societal implications ...

But there is more to this than that. There is a public 'gut feeling' (not that it is wise to try to think with your gut!) and sound scientific evidence that insights about the physical basis of mind could have implications for things that we consider to be important in our society. In education for instance. There have been tremendous developments in understanding - physiologically, pharmacologically and in terms of the molecular biology - the concept of 'critical periods' in the learning processes of children. These postnatal periods are critical time periods for the establishment of certain behavioural functions. From animal experiments we are beginning to understand the activity of dependent self-organisation and fine-tuning processes which happen in the brain during these critical periods. We also know that the neural substrates of the behavioural functions develop only if the nervous system receives certain environmental inputs during these critical periods. This gives us the beginning of an explanation why something like second-language learning later on in life is so different from the kind of language learning early on in life. Out of this kind of understanding come suggestions that we introduce second language learning in a better way in the first years of primary school.

The knowledge which is gathered by cognitive neuroscientists can also be applied to fields other than education. Management of certain mental health problems and age-associated cognitive deterioration are just a few which can be mentioned. I will not dwell on those, but these are other examples on which cognitive neuroscience might have an impact.

... and opportunities for interaction

Lastly, there are many opportunities for interaction between neuroscientists and computer scientists. During the course of the last 18 months the UK government, through its Foresight Cognitive Systems Project, is looking at the possibility whether useful ideas about the design of the emerging personal digital environments might come out of neurosciences, neurophysiology and psychology.

The cognitive neuroscientist

What is it that cognitive neuroscientists actually do? Cognitive neuroscience comprises research groups which may be categorised broadly into three groups. The first group comprises experimental scientists doing human neuropsychology and non-invasive brain-imaging using positron emission tomography [→ glossary] (PET), functional magnetic resonance imaging [→ glossary] (fMRI) and electro- and magnetoencephalography [→ glossary] (EEG and MEG) techniques to look at regions of activation in the brain during interesting mental tasks. And there is a growing interest on the part of the imaging research community in looking into aspects of social cognition.

Others are doing invasive experiments with animals, including non-human primates and other largely vertebrate species, including transgenic animals [→ glossary]. They use single-unit recording and various lesion techniques. There is a growing intolerance in Europe and perhaps in other countries as well as regards the use of non-human primates. That intolerance presents a challenge. If we are to get rid of the research on non-human primates for ethical reasons, we will be left with a huge gap between the work, which is done on rodents, and the application of this research to humans. We would have to live with that gap and with all of the implications of that.

The third group of cognitive neuroscientists is computational scientists doing modelling and theoretical work. And, of course, there are a few polymaths who do all of these three.

The collective aim of the enterprise - which is well represented in Europe - is to understand the organisational principles of information processing (e.g. of perception, attention, memory, language and action) by mapping these processes onto the regional and local-circuit networks of the brain and by understanding better the associated activity of individual neurons.

Working and long-term memory

I would like to give you one example with reference to the topic on which my research focuses, namely memory. Memory helps to define our individuality. It is through learning and memory that we acquire the various skills that we all need in our lives. Memories can be good things, but they can also be things which in certain circumstances it is best to get rid off.

One of the great discoveries of the last twenty years has been a refinement of our conceptual understanding of the organisation of memory. Two decades ago we thought of memory as just a unitary entity. The pharmaceutical industry was simply looking for drugs that would boost memory. Now we have a much more sophisticated understanding of the neuropsychological organisation of memory. We know much better where it maps on the anatomical structures of the brain and how the anatomical structures interact. Because of this increased knowledge, the memory system has grown out to become a showcase example of how to explain to the public how the brain works.

There are two kinds of memory: long-term memory and the working or short-term memory. The working memory is the kind of conscious memory through which we do things like mental arithmetic, remembering telephone numbers or as young children do, use it extensively for learning languages. There have been enormous contributions to the understanding of this memory by experimental psychologists such as Alan Baddeley. He is one of the architects of our developing understanding of the concept of working memory. Many of the insights into the organisation of working memory came from patients with brain damage, but more recently our knowledge has expanded largely because of the remarkable evolutions in brain imaging. Through work that has been done in a number of different locations around Europe, particularly by The Welcome Functional Imaging Laboratory, we begin to get an understanding of the activation in a whole variety of different brain areas, not just the frontal lobe. Many people still associate the working memory just with the frontal lobe, but that turns out to be an old concept. We have now good reasons to believe that working memory actually involves a constellation of areas which interact with another one. Many people still think in a very phrenological way about a specific area for this or a specific area for that. I choose this example specifically to make the point that a single entity like working memory consists of a family of processes, circuits and regions which interact and which are constantly changing their effective connectivity in order to mediate this particular process.

What about long-term memory? Long-term memory is also divided up into different bits and pieces and at the moment there are lots of taxonomic wars between scientists. Actually there is a major division between those of us on this side of the Atlantic and scientists in the US who tend to take a different framework for this. Endel Tulving, an Estonian who works in Toronto, has developed a framework which in my opinion is interesting. According to Tulving, memory is a cascade of systems in which we represent things first of all perceptually, then there are links to associative codes in semantic memory and finally there are links made with episodic memory. By episodic memory we mean specific moments in our lives, whether it be mundane things like what we had for breakfast this morning or more specific moments in our lives like getting married or our children being born.

There are lots of very exciting experiments going on in trying to capture where and how this all is happening in the brain. Some of these experiments have caught public attention like the one by Maguire who does really imaginative work. She took a group of London taxi drivers and put them in a PET-scanner [→ glossary] and asked them to imagine that they were driving from Grosvenor Square to Elephant and Castle (both locations are in the centre of London). In their minds the drivers would be navigating and imagining the drive probably by recalling images from the streets, traffic lights, landmarks etc. The really clever thing about her experiment is that Maguire asked the same taxi drivers to do a second mental task. They had to imagine a series of landmarks: the pyramids, the Eiffel Tower, the Empire State Building and so on. In the last task, the drivers would be imagining these landmarks but this time they would not have to navigate. Maguire was able to subtract the brain activity patterns of the landmark visualisation exercise from the London navigation/visualisation task. She found out that a particular area of the right parahippocampal gyrus lit up, linking that specific brain area to navigation. So these very ingenious and clever experiments are being taught nowadays by cognitive neuroscientists.

Brain images vs. single-cell recordings

However, there are sceptics who have a lot of criticism of these kinds of brain-imaging studies. They warn against an over-interpretation of the images. I agree partially with them. We have to think very carefully about how we interpret data from brain images. Functional MR images result from a haemodynamic signal and that is not a direct measure of brain activity. It is an indirect signal which is most likely correlated to neural activity, but it is not a direct proof of neural activity. Therefore, it is important to mention that some very exciting developments are happening in animal work with respect to imaging. These are taking advantage from other levels of analysis, particularly the molecular level. I would like to illustrate this point with one set of experiments from Tobias Bonhoeffer from the Max Planck Institute in Martinsried near Munich (Germany). Bonhoeffer is working with transgenic mice [→ glossary] which express a green fluorescent protein in about one in a hundred neurons. The Bonhoeffer mice are the modern counterpart of the Golgi's staining method for neurons which was used late in the 19th and early in the 20th century by the neuroscientists Camillo Golgi and Santiago Ramon y Cajal. This staining method lights up about one in a hundred neurons. The advantage of the Bonhoeffer mice is that they are alive when their neurons are expressing this fluorescent protein. The staining method of Golgi can be applied only on dissected brain tissues. Bonhoeffer was able to make optical sections through the brain of living mice using a two-photon confocal microscope. What you see are individual neurons in a living brain in action. That is quite exciting. I keep challenging my colleagues as to how quickly they can develop non-invasive human brain imaging which could take us down to the level of neurons. If we were able to do that, it would boost our knowledge of the living brain at the cellular and molecular level.

Animal models

I think that animal work can teach us a lot about the brain. Especially about the ageing brain. I agree with Susan Greenfield that, in an ageing society like West European society, we have to put the emphasis on the healthy brain and the opportunities for older people. But an ageing population also holds dangers because of a number of age-associated diseases, such as Alzheimer's disease. And here I do part company slightly with Susan. I agree entirely with her that the genotype-phenotype interaction is a much more complicated one than many people have been led to believe. Many science journalists are a little irresponsible when they talk and write about a gene for this and a gene for that. We definitely have to move beyond that. But in my opinion animal models are a great step forward in understanding the processes, which happen in ageing and deteriorating brains. The animal models can help us to dissect out the different aspects of some of these pathological processes. We should look at the animal models, and especially the transgenics, as furry test tubes. A furry test tube is an animal which is expressing a mutant gene of some kind and which might simply represent one aspect of a very complicated disease. Take the example of the PDAPP-mouse made by the Irish biotech Elan Pharmaceutical. This mouse over-expresses a mutant form of the amyloid precursor protein and as that mouse gets older it starts to lay down in the brain deposits of this protein in the form of plaques. This mouse allows us to dissect out the amyloid pathology in Alzheimer's disease[2].

We had a great debate recently in Britain about a primate centre and people were led to believe that primates are the good way forward to look at Alzheimer's disease. I do not agree with that view. I think mice are a great way forward for looking at neurodegeneration, even though mice live only two years. But you can engineer mice according to the questions that you want to solve, like what is the specific role of beta-amyloid, what is the specific role of presenilin. I agree that it is out of context of the whole complexity of the human brain, but it is a start. It is a way of trying to take things forward.

Deal with public understanding

So in our laboratory we have been doing some cognitive testing on these PDAPP mice. We are collaborating with the pharmaceutical company Eli Lilly to look at the impact of a drug, which interacts with the presenilins and the amyloid deposition, on the cognitive abilities of these mice.

Our work has attracted publicity in Scotland. Because of all the various problems in Britain with animal experiments I was a bit nervous when our work came out in the lay press. But I was amazed because my neighbours came up to me and said 'Why isn't there more of this work going on in Britain' and that was a delight. If you present the work in an appropriate context, you can actually have an intelligent discussion about the value of this kind of work.

Cognitive and computer scientists

To close my contribution, I was planning to talk about the recent UK Foresight Project on Cognitive Systems. The project has to do with the interaction between neuroscientists and computer scientists and it includes discussion of the social aspects of cognition and the likely economic impact of cognitive systems research. Unfortunately, my time is limited and I cannot go further into the aims and outcomes of this project. But I

2 More details on the involvement of beta-amyloid and presenilin in Alzheimer's disease can be found in the presentation of Professor Christine Van Broeckhoven.

want to give one example on how cognitive neuroscience and engineers can join forces to solve day-to-day problems in the digital society we will all be living in tomorrow.

Engineers and computer scientists have approached the task of automatic speech recognition using techniques which are called hidden Markov models. They take and transcribe lots of spoken vocabulary and by using machine-learning algorithms they try to link one to the other. That is the basis of the current speech recognition programmes. These programs were developed in the 1990's and they work reasonably well for speaker-independent dictation. Now it turns out that these programmes are not good at all in recognising speech in something ordinary like telephone conversations. To improve speech recognition in these kinds of applications they need some of the knowledge of psychology which can be brought in by the cognitive neurosciences. The speech-recognition programmes need to work out the contextual facts like 'what is it that one speaker knows and another speaker knows and how does that relate to the conversation'. It is the two-person business that Professor Jouvent was talking about earlier. Until the programme has that type of information it simply will not be able to recognize and understand speech in day-to-day conversation.

I hope I was able to give you a glimpse of what current cognitive neuroscience is about and what the tools are that neuroscientists use these days.

Discussion

SUSAN GREENFIELD: I do not want to go too deep into technical questions, I think we can discuss them among ourselves, but I want to make a point about animal models. Earlier in my talk I was not saying that the problem with animal models is that they are 'out of context'. I rather wanted to emphasise necessity and sufficiency. Just because something is there in a certain brain disease, it does not mean that it is necessary or sufficient for the neuronal damage in that disease. My own view is that there are also animal models that don't have histopathological markers like plaques but nevertheless have cognitive deficits. For example, models with an over-expression of acetylcholinesterase. My point is: as long as it is unclear what the primary cause is, it is very hard to model that cause. And I think the jury is still out as to what the primary cause of Alzheimer's disease is.

RICHARD MORRIS: I agree with you on that, although it is hard to have this discussion without getting very technical. But the problem is how could you ever work out the primary cause of Alzheimer just from research on humans. In humans the only pathology which one can study is post mortem end-stage pathology. With animal models, you can at least study some aspects, which are part of the whole pathology, in very early stages. With the models, initial emphasis was on plaques, but that has changed tremendously. The emphasis is now on over-expression of beta-mayloid or on what is happening when you knock out presenilin. So the focus of attention has moved through that biological work on transgenics. These mouse models become players that enable us to move forward.

PETER RAEYMAEKERS: Can you explain what the cellular or even molecular correlate is of memory and cognition. Allow me to give an example: I visited this meeting room about a week ago. Today, when I came back in this room, I recognised it. What is it at the molecular level that makes me recognise this room again and even notice that the seats and the tables are arranged differently?

RICHARD MORRIS: Recognition is a very nice example of memory. If you look at brain-imaging studies, what several have shown as regards recognition-memory is that areas of the perirhinal cortex in the mediotemporal lobe are activated specifically. So we know at the gross level which regions are involved. If you do invasive experiments on individual brain cells in primates that are trained to do recognition, likewise one sees changes in the firing pattern of cells in that area and less in other areas. The next step is to look at gene activation. There are genes, so called immediate early genes, which are activated in cells in these areas. So in conclusion, we are beginning to get the picture of the anatomical region, of the differences in firing rate in the cells in these regions and of the molecular players in these cells, even down to the level of activation of immediate early genes. So we begin to put together the bits and pieces of recognition and of memory.

BERNARD REBER: You have used for brain science the metaphor of the new astronomy. To have an indication of terms, time and process, where are we at the moment with brain science in relation to the history of astronomy? Are we in the universe of the Ptolemaic period, Galileo or Newton?

RICHARD MORRIS: Although it is a very interesting question, I find it very difficult to answer it globally. Certain areas of brain research are extremely advanced, others are less advanced. It is commonplace for people to say 'look we do not understand anything about the brain'. I get irritated when I hear that. In the last hundred years researchers have worked very hard and we do know a lot about the brain. But I agree that there are certain areas where our knowledge remains very primitive. So, one cannot make a statement about the status of brain science as a whole and draw analogies in precisely the way you suggested.

JÖRG NAUMANN: Could it be possible within five or ten years to enhance the capacity of memory or destroy some parts of memory?

RICHARD MORRIS: It may well be possible to do so. Of course it opens up all sorts of issues on which Steven Rose will want to talk about tomorrow. Like when would memory enhancement be an appropriate thing to do? I can imagine cases, even outside pathology and disease, in which the use of memory enhancement would be quite beneficial. What about people who have some element of benign forgetfulness but who want to keep their job? Wouldn't some kind of pharmaceutical enhancement, which restores their memory capacity so that they can stay on the job for longer, be valuable? Wouldn't that be an entirely beneficial kind of use? But whether it would be reasonable for children to take such drugs prior to exams is another matter. Would we look upon that as like the doping problems in sports?

But if you ask me whether it be technically possible in ten years? Yes, I think it will. There is a great deal of interest in the pharmaceutical industry in both mild cognitive impairment and more serious problems. We are making progress in understanding some of these diseases and some of the normal mechanisms through which memory is encoded, like plasticity.

KARIN RONDIA: What about erasing memory? The issue is coming up in a few American movies which have not yet been on show in Europe.

RICHARD MORRIS: It is an important topic, but it is not one I know a lot about. So I am not going to answer that question, I will leave that to others.

STEVEN ROSE: There are indeed speculations about the possibility of memory erasure. Over the last two or three years it has been demonstrated in animal models that when an animal is reminded of a previous experience, the memory becomes labile once more and can be erased by injecting metabolic inhibitors such as protein synthesis inhibitors or neurotransmitter inhibitors. This has led some American colleagues to speculate that this might be a way of treating post-traumatic stress disorder. I happened to see last night the film that Karin was referring to. It is a Kauffman film called 'Eternal Sunshine of the Spotless Mind'. An unhappy love affair is erased as a result of some kind of transcranial magnetic stimulation [→ glossary]. I am not going to comment on the content and the quality of the film, but at least the ethical issues behind it are extremely interesting.

KARIN RONDIA: Is anyone aware of any real life application of transcranial magnetic stimulation on the therapeutic side?

WRYE SENTENTIA: Transcranial magnetic stimulation is being used to treat post-traumatic stress disorder in clinical studies. More interestingly, it is also being used to find the 'God spot' in the brain. For that application, it is employed in conjunction with fMRI studies of monks and religious people who are in meditation. By stimulating certain brain regions researchers try to evoke that spiritual state.

PAOLO LUCIO MORSELLI: Transcranial magnetic stimulation is applied today in depression and bipolar disorder. With some success.

STEVEN ROSE: I also want to make the point that there is very heavy investment in research in the area of transcranial magnetic stimulation in the US. This research is funded by DARPA, the Defence Advanced Research Projects Agency, the research department of the Pentagon and the US army. In this particular context this Agency is not interested in therapeutics, I would say.

Brain imaging, the present, the future and societal consequences

BY PROFESSOR MARK VAN BUCHEM

Mark van Buchem is Professor of Neuroradiology at Leiden University Medical Centre in the Netherlands. His research group has recently discovered that migraine leads to small infarcts in the cerebellum. According to Professor van Buchem brains differ among people as much as individual faces. The art of the neuroradiologist is therefore to neglect normal variance and detect subtle deviations in the brain. In his presentation, he tells how he practices his art.

For a physician like me it is quite unusual to be challenged to talk about the present, the future and the societal consequences of brain imaging, and all within a very short time frame.

The history of brain-imaging attempts goes back to the good old X-ray days. But unfortunately, if you X-ray a head, you do not see the brain. First of all, there is the skull which absorbs most of the X-rays. Secondly, the X-ray absorbance capacity of the different brain components is very similar. As a result, the radiologist does not get enough contrast between the different parts of the brain to make them differentially visible. That is why other techniques have been developed. Back in the nineteen seventies, the first real images of living brains came from computed tomography (CT) [→ glossary]. CT has nowadays become much more sophisticated than in the early days. One can make all kinds of interesting and informative reconstructions like giving different colours to blood vessels etc.

Magnetic signals

Another interesting imaging technique is PET scanning [→ glossary], but what has really become our tool of choice as radiologists is magnetic resonance imaging (MRI) [→ glossary]. MRI is a scanning technique which uses not radioactivity but a high magnetic field. To get images, we put patients in a tubular scanner surrounded by strong electromagnets and all kind of antennas and sensors. It is the interplay between a magnetic field and radio frequency pulses that gives rise to these beautiful images that we take for granted nowadays. The MR images that are usually shown are not the 'raw' magnetic signals that are radiated by the molecules in the brain and that are measured by the sensors, but are mathematical transformations performed on those raw data.

The whole art of MR imaging is to get contrast and to make sure that one gets differences in signal intensity between different structures of the brain. Without contrast, there is no image. Therefore, the absolute intensity of the signal, the grey level per pixel, is completely irrelevant in most clinical applications of MRI as long as there are differences in signal intensity between structures.

The image shop

We can generate all kind of images, like T1-weighted images, which are in a way the workhorse of the radiologist. These images show very nicely the anatomy of the brain with the grey and the white matter. The T2-weighted images are more sensitive for detecting disease. But there are many more: proton density-weighted images, diffusion-weighted images, techniques that focus on the detection of flowing blood (this method allows a three-dimensional image of the whole brain vasculature to be reconstructed) and many more. They all tell you something different about the brain. Each of them is based on different phenomena in the brain.

But for neuroscientists there are even more interesting applications, like a new technique called diffusion tensor-weighted imaging. This allows detection of the preferential direction of the white-matter tracks. The technique permits detection of the internal architecture of the brain's white matter, in other words the wiring of the brain.

Then there is also MR spectroscopy, a method whereby we screen metabolism in the brain. This gives you more functional information. Based on that metabolic information, one can get an idea of the different biochemical reactions that are taking place at different locations in the brain. An example of this is a lactate image of the brain. High concentrations of lactate in the brain are an indication that non-aerobic processes are taking place in the case, for example, of an infarct.

Radiologists use these images to detect all kinds of pathology, from tumours to multiple sclerosis. Moreover, these techniques are not only useful after birth, but also before birth, and this is something we do in Leiden on a weekly basis.

But MRI is more than nice pictures and images. There is also quantitative imaging. With quantitative images the grey scale in the pixel is meaningful, by way of distinction with qualitative imaging where the contrast between different structures is most important. With quantitative techniques like magnetisation transfer-imaging, we measure the concentration of myelin in the brain. This often allows us to make better correlations between the images and the clinical status of a patient. This also allows abnormalities to be detected in qualitative imaging, such as MR plaques or brain tumours which can be quantified. For this to be done correctly, image processing techniques are required.

Different faces, different brains

For radiologists individuals brains differ as much as faces. Because of the tremendous variations between the brains of normal individuals, it is hard for radiologists to detect subtle, pathological differences. The only way to deal with that is through a process of normalisation. This processing of the image is performed by the computer. We sort of push back the image of the brain in a fixed mould so that each part of the brain is placed in the same position. That increases enormously the similarities between normal brains.

This type of normalisation is very useful in research. If you take, for example, the brain images of a hundred normal people and if you normalise all of them, you get something like an 'average normal brain'. In such an averaged brain, every pixel contains information regarding the structure of those hundred brains. That allows you to compare brains from populations of normal people with people who are affected by a disorder, say schizophrenia. Comparing the normalised brains of both groups allows you to infer on a statistical basis how the brains of schizophrenic people differ from normal brains. If you look at all the individual brains, you will never discover that there is any difference at all. So this is a powerful tool, but it only makes sense if you look at groups of individuals.

Another example I want to give is a study by Arthur W. Toga's group at the Laboratory of Neuro-Imaging at the University of California in Los Angeles (USA). They looked for similarities between normalised brains from populations of unrelated persons, dizygotic twins and monozygotic twins. We all know from experience that the faces of monozygotic twins are much more similar than the faces of dizygotic twins which in turn look much more alike than the faces of unrelated people. The same is true for the outer structure of the brain. The structural similarity between the couples of monozygotic twins is much higher than the similarities between dizygotic twins and unrelated people. This shows that the structure of the brain is in large part genetically determined. Of course this does not say anything at the functional or behavioural level.

The mapped neuron

Lastly there is also molecular imaging. It is basically *in vivo* diagnosis of complex processes using specific agents. With molecular imaging, we try to detect abnormalities at the cellular level. With the earlier discussed qualitative and quantitative methods we can detect only abnormalities of large groups of cells. In molecular imaging the trick is to have a carrier, for example a monoclonal antibody to which something is bound that can be detected with MR, like a metal atom. You inject that complex into the body and you can detect very specific molecular structures that can be picked up by the monoclonal antibody. If your antibody is binding to fibrin, than you can detect in blood vessels where fibrin is present. That could be, for example, an indication of a small blood clot.

The future

What will the future bring? Qualitative imaging is still on the rise: resolution is improving and the time needed to collect the image is becoming shorter. Quantitative imaging at present is used only for research but will find its way to the clinic. Molecular imaging will be applied to a greater extent. Imaging-process techniques will become even more powerful.

Society

The societal issues? Early detection is an issue. We shall be able to detect diseases much earlier than at present. Maybe even at a stage when the pathology is present only at the cellular level and no symptoms are present in the patient. The consequences of this are enormous. Are we going to set up population screening programmes for brain diseases? At least in the Netherlands population-based screenings are for the moment not accepted. But it could become an issue, certainly when there are early treatment options for certain disorders. It could also be that insurance companies will want to subject individuals to a similar screening.

The second issue, which is that of increasing recognition of structural substrates for functional differences, is at least even more pressing. I referred to schizophrenia patients earlier. A lot of psychiatric patients feel guilty because of their disorder. If they know that something structural is abnormal in their brain, the guilt problem might disappear. This might also have an influence on the therapeutic treatment of psychiatric disorders: if they know they have a structural abnormality, people might opt to a greater extent for a biological treatment instead of psychotherapy. But insurance companies and employers also might want to screen people to find out whether they are more prone to psychiatric and neurological disorders.

I would like to conclude my talk with these quick societal remarks. I thank you for your attention.

Discussion

BERNARD REBER: I have a more technical question. What do you do with the data that you gather from patients and subjects in terms of security, privacy and consent?

MARK VAN BUCHEM: First of all, there are the patients we scan for diagnostic purposes. The images and the data from those patients are stored and linked to their medical files. We might have to retrieve these at a later stage to follow the course of the disorder by comparing images from different time points. For our research, it depends. If we are doing clinical research, the information is stored in our database. If we do population-based studies, the information is stored anonymously.

SUSAN GREENFIELD: I want to pick up on a very fascinating point you raised when you explained the societal implication of people no longer feeling responsible because they can say 'my brain has predisposed me to behaving like this'. It is an interesting issue at the crossroad between neuroscience and law. But how can you be so sure that the structure precedes the function when we know from research that activity shapes the brain as well? One cannot say that just because your brain functions in a certain way, one can predict certain behaviour, while it could be the other way around? Is it not a two-way system?

MARK VAN BUCHEM: Absolutely. You raised a very good point and we are not sure yet. But if structure precedes function, as I suggested, it may have severe societal consequences. But I agree that we are not there yet.

STEVEN ROSE: There are already arguments in Great Britain and in the US that you could use proactive brain screening of this sort which precedes the event. For example, to detect potential psychopathy. Might it then be defendable to detain people in advance before they have committed any criminal act at all? That is a serious debating issue in Britain at the moment.

ANDREAS ROEPSTORFF: I agree with Susan Greenfield that there is a big discussion as to whether function precedes structure or structure precedes function. I would interpret the images which show that the brains of monozygotic twins are much more similar than completely different as Prof. van Buchem has done during his presentation. If you had shown the faces of the twins, they would have looked much more similar than the brains. If you look at the images of the brain, one can observe that the similarities are located mainly in the primary sensory and primary motor areas. In all other parts of the brain, there are very important differences. You can take this picture just as much as an index that, although monozygotic twins have an identical genetic make-up and even though their faces would look very similar, when you look at the more interesting parts of their brains the parts which have to do with cognition, memory, learning and other things, these parts differ a lot. That observation points to two issues. First of all that structure does not exclusively precede function; function and activity seem to shape structure as well. But it points also to the fact that the interpretation of these brain images is extremely difficult and that the same image can be put into a very different discussion and suddenly show something completely differently. I would take this picture to indicate that, although the genetic make-up is the same, when you look at the cognitively interesting areas of the brain, the two twins are actually very different from each other.

MARK VAN BUCHEM: I agree with you. The only thing that this study shows is that there are areas in the brain that are genetically predisposed in terms of shape, and we are of course not talking about function here. But I also agree that there are recent studies which show that longitudinal mental tasks can change the shape of brain regions. So indeed it seems to be a two-way system.

Leo Zonneveld,
British Embassy,
The Hague,
The Netherlands

LEO ZONNEVELD: As far as MRI imaging is concerned, the quality of imaging is related to the strength of the magnetic field you allow. At this moment you can go to 7 tesla, which is the maximum a human being can endure. Is it possible to create images which are as good as we want them to be without increasing the magnetic field or the radio frequency indefinitely?

MARK VAN BUCHEM: In clinical practice 1.5-tesla is widely used, but increasingly 3-tesla machines are making their way to the clinic. The next step is certainly going to be the 7-tesla machine. Why are we developing these machines with high magnetic fields? There are two things: by increasing the field strength one can create an image with the same resolution in less time, or in the same amount of time one can create an image with superior resolution. So you can do quicker experiments or experiments with better resolution. Therefore, higher field strengths create options, which you cannot get with lower field strengths.

Neurosurgery

BY PROFESSOR JAN WILLEM BERKELBACH VAN DER SPRENKEL

Jan Willem Berkelbach van der Sprenkel is a neurosurgeon and Head of the Neurosurgery Department of the Utrecht University Medical Centre in the Netherlands. His main fields of interest are neurovascular research, neuronavigation and functional neurosurgery.

Preservation of function has nowadays become the key issue in brain surgery. For us as surgeons, it still remains amazing that you can take out big parts of the brain without really damaging any kind of function. In the early days, neurosurgery was mainly based on anatomical knowledge of the brain; now neurosurgeons base themselves both on structure and function. To be able to do that, surgeons are very dependent on imaging and functional recording techniques before and during the operation.

Improvements

What has changed neurosurgery in the last decades? Recent improvements in neurosurgery are due to three main factors. First, there are the developments in anaesthesia and post-operative caring techniques. I will not discuss these because they fall outside the scope of this workshop. Secondly, the use of the surgical microscope has had a tremendous impact on brain surgery. The microscope not only visually enlarges the operation field but also provides superior illumination. Both were necessary to make the development of microsurgery possible. The development of surgical endoscopes has provided further improvement of visualisation during surgery.

But maybe most importantly, there has been tremendous progress in diagnostic procedures. As was discussed in the previous talk by Mark van Buchem, MRI has become the golden standard for the diagnosis of brain diseases. But MRI did more. It has really allowed the development of neuronavigation during surgery.

Navigate with the scalpel

Neuronavigation uses pre-operative diagnostic images and takes them along during the operation. The navigation matches the 3-D volume images of the pre-operative scans of the patient's brain in the operating room. So even with a closed skull you know where to locate the pathology within millimetres. Simply put, before an operation is performed several images are made of the brain of the patient and stored in the computer. During the operation, the head of the patient is filmed by a camera and the location and position of the head are linked to the pre-operative MR images. All of this is shown on a screen in the operating room. By looking at the screen, the surgeon knows where the pathology is without even opening the brain. The aim of the whole enterprise is to take out as much pathology as possible without damaging the cognitive and other functional abilities of the patient. In order to use neuronavigation safely during surgery, brain movement should be monitored during surgery, and this is a technical problem that has not yet been entirely solved.

From a scientific and technical point of view, it is also interesting to develop ways of presenting this information to the surgeon while he is operating. Looking at visual information on a computer screen while you are operating is not always very practical. One has to think of other modalities as well, like auditory information or perhaps even tactile information.

Neurosurgeons have already been working for several years with this pre-operative imaging technology, although at different levels of complexity. But pre-operative imaging is not sufficient. During the surgical procedure, the anatomy of the patient's brain is changed. The pathology is taken out, parts of the brain are removed, and so on. So what we need is to image the changing anatomy and functional information during the operation itself.

Functional navigation

As I mentioned earlier, it is not only images and anatomy that count in navigation during surgery. The cortical and other brain-areas representation function is even more important for the surgeon. We cannot see on the outside of the individual brain where certain functions are located. So how can we find out in our individual patient on whom we are operating where exactly his brain functions are localised? In theory, we would be able to get that information from functional MRI investigations before the operation, but again that would not be sufficient. We need to confirm the localisation during surgery. It would be highly impractical to bring a patient during surgery to a MR scanner or to perform the surgery in a scanner. So the only practical way to solve this problem is to operate while the patient is conscious and let him do mental tasks or functional tests like speech-arrest experiments.
The only technical alternative to detecting changes during the operation for which I see some promise now and in the future is the use of ultrasound. It is a much more simple technique than MRI and is practicable during surgery.

The ultimate surgeon

The ultimate surgeon might eventually be some kind of robotic system. We are not talking about an independently working robot. It is more a restraining robot which you can programme in such a way that it restrains the movements of the surgeon. In case of a tumour resection, the robot would prevent the tip of the suction device from being moved outside the area of the tumour. The area of the tumour has been defined on the basis of the images taken pre-operatively and is stored in the computer which also keeps track of the movements of the suction device during the operation. This surgical technology can come into practice in the immediate to near future.

Deep brain stimulation

But nowadays, brain surgery is becoming more than tumour resection. The success of functional neurosurgery is increasing. For example neurosurgery on epileptic patients has become quite common and is performed all over the world now. Also, electrical stimulation techniques *in vivo* are changing functional neurosurgery more and more. In deep brain electrical stimulation [→ glossary], you put a wire with an electrode deep into the brain and let it make contact with groups of nerve cells. The electrode is connected to an implanted pacemaker which stimulates the nerve cells with mild electrical currents. This kind of deep brain stimulation has become a nearly routine procedure for patients with Parkinson's disease who have become resistant to drug therapy.

An example of deep brain stimulation is the patient who received an implant near the thalamus. [Professor Berkelbach van der Sprenkel shows an impressive video of a patient with the Gilles de la Tourette syndrome before and after implantation of the electrode.] As you can conclude from the video fragments, the effect of the stimulator on the life of this man has been tremendous. Until now, only a few patients with the Gilles de la Tourette syndrome have been transplanted by my colleague neurosurgeon, Veerle Vandewalle of the University Hospital in Maastricht (the Netherlands).
An advantage of deep brain stimulation is that one can easily vary the treatment (more or less of the stimulation intensity) and that it is completely reversible. As a result, one can do crossover studies with the patient as his own control. It is possible to turn off the stimulation and see what the effect is.

History of Psychosurgery

Why is deep brain stimulation of such interest for a workshop on the societal issues involved in brain science? Well, the field of deep brain stimulation is extending, and it is extending to psychosurgery. As we all know, the history of psychosurgery is a bit gloomy. The first widely advertised cases of psychosurgery were performed by the Portuguese neurologist Egas Moniz. Moniz made lesions on both sides of the frontal brain in patients with psychiatric disorders. He cut the nerve fibres which connect the frontal and prefrontal cortex to the deeper located thalamus. The thalamus is responsible for relaying sensory information to the cortex of the brain. Moniz suggested that an interruption of this pathway would inhibit repetitive thoughts, allowing a more normal life for the psychotic. He observed that the psychiatric patients on whom he performed this procedure became much more quiet and manageable. His technique, which became generally known as lobotomy, was considered to be as promising a development in medical progress as X-rays. Moniz therefore received the Nobel Prize in 1949.

Lobotomy was picked up by Walter Freeman and James Watts in the US. While Moniz was drilling holes through the skull, Freeman and Watts 'refined' the procedure by developing the famous transorbital approach. They literally pushed ice picks into the brain via the eye sockets. Between 1936 and 1956, approximately 60,000 of these procedures were performed in the US. The results were actually not as bad as one might think. About 40 percent of these patients improved. Two percent got worse and 4 percent died. But the main reason why this procedure was abandoned was because of the severe side effects, especially passivity and loss of emotions.

Nowadays, lobotomies are done only seldom. In the past few years, we treated some patients with very severe and untreatable obsessive compulsive disorder (OCD). The lesions done nowadays are thermal lesions. But personally, I think that with the new introduction of deep brain stimulation, this lesion technique will be a thing of the past. A publication in the journal Neurosurgery in June 2003 by the group of Bart Nuttin of the University of Leuven (Belgium) tells the story about five patients with OCD who had an electrode implanted. The results of this treatment seem to be very promising. With this technique, we have for the first time in psychosurgery the chance to help patients without damaging their brain in an irreversible way. Of course, there are still a number of ethical questions related to this kind of surgery, but I leave that discussion for tomorrow.

Discussion

KARIN RONDIA: What will be the limits of deep brain stimulation? Would you be able to stimulate every nucleus or every nerve-cell group in the brain?

JAN WILLEM BERKELBACH VAN DER SPRENKEL: One of the limiting factors right now is definitely the size of the electrode. The current electrodes are a few millimetres thick, compared to the size of the individual brain cell that is relatively thick. At the moment it is a challenge to construct thinner electrodes, which are still firm enough to be used. But in principle, I would say that it is possible to stimulate every region of the brain. Although one should be careful about side effects. These side effects should be mild if you hit your targeted brain area right in the middle with the electrode. In that case, you can use relatively small currents to stimulate the region. But if you are a little bit out of your way, you need stronger currents and you also stimulate other regions. That can induce unwanted side effects. So right now, we do not really know what the limits are.

PAOLO VALERIO: I was wondering whether we know enough about the functioning of our brain to use coarse methods like neurosurgery. In the example of the lobotomy, I don't agree that the side effects were only passivity and loss of emotion. The side effects were much more severe in large numbers of patients. Aren't we still too early with brain surgery?

JAN WILLEM BERKELBACH VAN DER SPRENKEL: I think that is exactly why neurosurgery deserves its place in workshops like these. In my opinion, we have to be very, very careful in doing these kinds of experiments. When we perform surgery we have to take a large number of arguments into account, especially when, during the operation, there is a risk of causing structural damage to the brain of the patient. If you don't inflict damage to the brain, as with the reversible technique of deep brain stimulation, there are lesser problems. But I agree that even then, there are still some ethical issues which should be discussed.

MARK VAN BUCHEM: How silent are silent areas in brain imaging? I have asked myself this question many times because neurosurgeons make the assumption that, if areas do not light up on a functional MR image, these areas have no function. But I guess that every region of the brain has a function. Only we don't have the right paradigms to demonstrate that that specific area indeed has a function.

JAN WILLEM BERKELBACH VAN DER SPRENKEL: I started my talk by saying that I am always surprised that you can take out large parts of the brain without causing any apparent functional loss. If we can miss these parts, why are they there? Intuitively, we feel that they should have a function, no? After all, I guess you might be right that surgeons often inflict subtle losses of function, simply because the gross functional tests they routinely carry out during surgery are not sensitive enough. When you bring cognitive neuroscience into the field of neurosurgery, you see indeed that the picture changes. Then indeed loss of some function is often present after surgery. But it is not the type of loss of of function that you spot immediately when you meet somebody or speak to someone for five minutes. But these could nevertheless be changes that are very important for patients' lives.

SUSAN GREENFIELD: Have people been looking at the long-term effects of deep brain stimulation? Together with the stimulation, I guess that all kinds of chemicals are being released by the stimulated neurons. What could be the long term effects of this? Secondly, could neuroplasticity play a role in the effects of deep brain stimulation?

JAN WILLEM BERKELBACH VAN DER SPRENKEL: Neuroplasticity plays a major role in brain recovery, not only in deep brain stimulation, but certainly also in other neurosurgical procedures. Answering your other question is more difficult. Results from current studies on deep brain stimulation mean that we know already something about the effects in the medium term. By leaving the stimulator on for three months and then turning it off for a while, we see that the relapse is nearly immediate, but often not complete. So for optimal treatment, the stimulation has to be continued for longer periods, probably even forever. Of course, one should interpret all these data with some caution, because the procedure was applied relative recently and only on five OCD patients. So as regards the real long-term effects, nobody knows yet.

Brain research deserves a better (funding) fate

BY PROFESSOR JES OLESEN

Professor Jes Olesen is a neurologist at the Glostrup University Hospital in Denmark. He is a specialist in headaches. Professor Olesen also fights for improving the image of the brain and brain research. In 1999 he organised in Denmark the very successful 'Year of the Brain' and in 2002 he founded the European Brain Council. Professor Olesen is the current president of the European Brain Council and his organisation devotes itself to uniting the different disciplines in brain science and to getting integrated brain research programmes on the European funding agenda.

I am very pleased to be here because in the European Brain Council we try to unite all good forces to promote brain research. Obviously, in the past people haven't been very successful in bringing together all the professional specialists who work on and with the brain. An indication of that is the complete disproportion between the importance of brain diseases and investment in brain research. Why is that?

Bad press

People are afraid of the brain. They are afraid of all the potential ethical issues that surround this whole field. There doesn't seem to be any problem about trying to do something better for the heart. But when you touch the brain ... then that is a completely different situation.

I think people are scared of the brain. We see only the bad things about it. We all think that our own brain doesn't work well enough. We feel our brain only when we have a headache. We think of brain disorders as incurable. The brain has become a synonym for all kinds of negative things. The brain has a bad image and a bad press. People do not think of the brain as the seat of affection, of being in love, even of the orgasm, they all happen in the brain.

European Brain Council

Let me briefly introduce the European Brain Council (EBC). The organisation is only two years old. The idea behind it is that we need a forum in which everyone who is interested in the brain and brain diseases can unite and work for a common cause. This field has been so fragmented in the past. Neurologists, psychiatrists, neurosurgeons and other professional groups each worked in their own corner. Moreover, there was not much coordination between patient groups and professionals. Basic brain scientists stood against clinical scientists and so on. As long as this field remains fragmented, there is no way that the science of the brain can get the prominence it deserves and takes its rightful place alongside cardiac research or cancer. So the aim of the EBC is primarily to promote brain research and do everything we can to improve the treatment and care of the people who suffer from brain diseases. And maybe I have to put it more explicitly by saying that by brain disorders we mean all disorders that affect the brain. Both neurological and psychiatric disorders are brain diseases. Some of them are due to psychological mechanisms, but they essentially take place in the brain.

The Brain Council is a European initiative. The members of the Council include pan-European bodies representing neurologists and psychiatrists, as well as patient federations in those respective fields, neurosurgeons, basic neuroscientists, and relevant areas of industry and commerce. The European Federation of Neurological Associations (EFNA) represents Europe-wide organisations of patients with neurological illnesses. The Global Alliance for Mental Illness Advocacy Networks (GAMIAN-Europe) represents associations of psychiatric patients. The European Federation of Neurological Societies (EFNS) represents national societies of neurologists throughout Europe and the European College of Neuropsychopharmacology (ECNP) is a Europe-wide organisation of psychiatrists and pharmacologists and other basic scientists. And finally, the European Association of Neurosurgical Societies (EANS) is a federation of national societies of neurosurgeons. EBC seeks to have industry representation as well, but in most cases, industry does not yet have subdivisions dealing specifically with the brain at a European level. Until that happens, we accept individual companies as representative of their industry. For example, AOK (Allgemeine Ortskrankenkasse) as the biggest German health insurance company - represented by AOK Schleswig-Holstein - is an EBC member and representative of the European insurance industry. But since most of our member organisations are umbrella organisations with national societies under them, one could say that the EBC represents tens of thousands of scientists and millions of patients.

By the very nature of our organisation, we are actually working very much in the spirit of the commission for research in Brussels which is so keen on breaking down the barriers and having people work together across the borders of national countries and across the borders of disciplines.

Burden of disease

We are now trying to develop a white paper on a future brain research programme for Europe. We shall be starting this work hopefully this summer but we are limited mainly by our administrative capacity and our funding. It is always very difficult when you start new initiatives. You have lots of ideas and you want to get new things started, but there is always limited capacity.

Another initiative of the EBC, but one which is currently being developed and which will be finished within a year from now is a project on the cost of brain disorders in Europe. This project analyses all the major neurological and psychiatric diseases. We have researchers looking at the epidemiology of the disorders and a group of health economists who will be analysing the economic consequences of these diseases based on the literature and data from the epidemiology. At the end of the project, we shall know how much all these disorders cost Europe, something that is currently not known.

One thing that we have been doing already is to look at a study conducted by the WHO and which is called 'The Global Burden of Disease' (GBD). This study looks at the burden of disease in general and on a global scale. The initial study in 1990 provided new measures that are now becoming universally accepted and have been used also in a repeat study, the GBD 2000. The key parameter of the study is disability-adjusted life years (DALY), which is the sum of years of life lost (YLL) due to premature death and years of life lived with disability (YLD). In the report that we have drawn up, data from the GBD 2000 study and the World Health Report 2001 on brain diseases were extracted for the territory of Europe. This territory corresponds roughly to the member countries of the European Federation of Neurological Societies. Throughout Europe, 23% of the years of healthy life is lost and 50% of YLD are caused by brain diseases. Regarding the key summary measure of lost health i.e. DALY, 35% of the total is due to brain diseases. We all know that the problem is increasing because the population in Europe is getting older and mortality from other causes such as heart diseases and cancer is expected to decrease.

The fact that approximately one-third of the disease burden is due to brain diseases should have an impact on the allocation of resources among teaching, research, health care and prevention. But guess what: if we compare these burden figures with the dispersion of budgets for health research in Europe, we find that only 8% of the health research budget goes to brain research.

Although other factors are also of importance, it seems reasonable that one-third of life science funding should go to basic and clinical neuroscience. In addition, the allocation of resources to prevention, diagnosis and treatment of brain diseases should be increased to approach, at least, one-third of health care expenditure. With the present data on hand, neurologists, neurosurgeons, psychiatrists, patient organizations and basic neuroscientists are in a better position to increase the focus on the brain.

Opportunities

But we should not look at the brain as just a source of burden and costs. Brain science is an area of opportunity. There is now an explosion going on of new insights into the brain. Of course, this new knowledge brings with it a number of ethical issues. Issues which were never discussed before: coma and brain death, patients who are unable to give consent to treatment because they are psychotic or cannot speak, stem cell therapy, curing drugs which are abused for enhancement ... and so on. Unfortunately I do not really have the time to discuss these issues now in detail but we will have a chance to debate them tomorrow.

I would like to end with the following recommendation. I hope that the ECD initiative will help in our struggle to get brain science on the table of the public and the decision-makers. But I also want to sound a warning: make sure that this initiative does not end up with a lot of worries: worries about this, worries about that, my God if you stimulate these electrodes in the brain, what might happen and so on. I hope that the problems which will be discussed by the public are being matched against the significant importance of brain diseases and the terrible impact of these diseases on millions of individuals, on their families and on society as a whole. So make sure that your initiative becomes a positive enterprise. Analyse the problems, but reach solutions and try to ensure that we move forward.

Discussion

RICHARD MORRIS: The European Brain Council is a wonderful initiative, but in your Council is it easy to have a dialogue amongst these different professional groups. We might be in a situation that the public would imagine that neurologists would find it very easy to talk to psychiatrists and basic scientists. But is that really the case?

JES OLESEN: In the Council this is not really a problem, most likely because all our members are convinced that optimal communication among ourselves is necessary to get things done. But I agree that in general there is far too little communication. So the hope is that by having some sort of umbrella organisation in which the different groups cooperate and collaborate, the spirit of communication will gradually spread down the system. Fortunately, we already see hopeful initiatives. In many hospitals, they now have centres for clinical neuroscience which includes neurology, psychiatry, neurosurgery, neurophysiology and so on. But we also think that it is very important that patient organisations closely cooperate with the professionals and that they work hand in hand.

Session 2

| Plenary discussion concerning the possible role of brain science in society and future developments in brain science |

MODERATED BY JAN STAMAN

JAN STAMAN: Perhaps the best way to start is to get a feeling, an impression of what was said earlier tonight. Who wants to start? Miss Faber, what is your feeling?

BERIT FABER: I think this day is the start of a very important initiative. With what we heard today, we are able to get an idea of developments in the area of brain research. But it will be very important to link the specific medical, technological and scientific knowledge properly to the ethical issues. For me, the explosion of new information was a revolution. But you should ask the other scientists whether what they heard was really a new revelation for their domains. For me as an outsider, as a non-brain scientist, it was a revelation.

JAN STAMAN: But do you really think that these scientists make a better world? Do they help society?

BERIT FABER: I am convinced that all the scientists we heard today have 'pure hearts'. They want to do good for society. But I also believe that it is important that they enter into dialogue with the public and create an understanding with the public. But they should have faith in true dialogue, two-way communication. They should commit themselves to going beyond pure educational dialogue in which knowledge is only transferred from the doctors and scientists to the public.

JAN STAMAN: Jörg Naumann, what struck you during the first hours of this workshop? Where do you see the promise of this brain science for society?

JÖRG NAUMANN: I see many promises. But what struck me mostly was how science can often be an art. If you look at the presentation of all these brain images ... they are pure art. Probably this building, being formerly designated to bring science and art together, is the right place to present these very artistic images. Furthermore, I do not believe that the brain has such a bad image as Professor Olesen has said. OK, maybe the brain is taboo, but I don't feel fear about the brain.

The image of psychiatry

RINIE VAN EST: Brain imaging seems to be very much a driving technology behind brain science. For genetics, that is less clear. There seems to be a discussion whether genetics is going to help us much in fields like psychiatry or neurological diseases. Furthermore, what has also struck me is that psychiatry is becoming a 'real science'. Why do I say a 'real science'? Well, as a lay person, when I thought of psychiatry in the past, I thought of Freud and Jung and at the same time I never really believed it. But today, after hearing these presentations, I have the feeling that there is real progress now. That we are getting there.

JAN STAMAN: Isn't that reductionism? We believe in psychiatry because we have an image of the mind now?

RINIE VAN EST: No, I do not think it has to do with reductionism. It has more to do with proving things and making them visible.

The image of the brain

JES OLESEN: Still, I want to make the point once again that the brain has a bad image. How can you otherwise explain the discrepancy between the burden that brain diseases cause and the amount of funding that we invest in brain research? The brain definitely has a marketing problem.

JAN STAMAN: Couldn't it be that the brain has not been discovered by the public yet? That the public is not really afraid, but that they just haven't seen the importance of the brain and brain diseases yet?

JES OLESEN: When you talk about brain disease, you actually talk about a new concept, a new term, which has been coined by only a few people. In the past, there has always been the distinction between diseases for neurologists and diseases for psychiatrists. That is ridiculous because nobody 'owns' those diseases. The distinction between the two is absolutely superficial. It is ridiculous to say, for example, that dementia is a neurological disease. The psychiatrists will say that it is a psychiatric disease and for the general practitioner it is a general disease. Diseases can't be characterised by professional speciality. They can be characterised only by the place where they are happening. And for both neurological and psychiatric disorders that place is the brain.

JAN WILLEM BERKELBACH VAN DER SPRENKEL: I think one of the reasons why the brain has a bad image is because people have the perception that neurological and psychiatric diseases cannot be treated properly, let alone cured. When you look at funding for cerebrovascular disease in the Netherlands, it has always been relatively easy to increase funding for heart diseases, but for strokes it has been much more difficult. In my opinion this discrepancy has a lot to do with the general thinking that there is nothing that can be done on stroke. If you have had a stroke, then that's it. It has happened to you and you die from it or you have to live with the consequences of it, but you cannot do anything about it anymore. If there is one thing that we have to teach the public, than it is that plasticity of the brain is very important.

PAOLO LUCIO MORSELLI: I think the reason why brain science gets so little support is because brain diseases are still very stigmatising. There is a lot of discrimination and prejudice surrounding brain disorders. We should enlarge the information stream and educate people, politicians and decision-makers, telling them that it is possible to cure many of these diseases if brain research is properly supported. It has been done in other fields like cancer and cardiovascular diseases. For those disorders, there has been a lot of information, publicity and marketing.

JAN STAMAN: Do you mean that this has not happened as regards brain science? Haven't brain scientists, neurologists, psychiatrists and neurosurgeons not yet said to the public 'listen, our science and technology are growing so tremendously that we really can help you'. Has this message not yet been communicated?

PAOLO LUCIO MORSELLI: Maybe that message has been put forward already, but probably not enough and not in the right way. We are living in an anti-science society where the industry, and especially the big pharmaceutical industry, is seen as representing the devil. On top of that, psychiatry suffers from many ideological divisions. We should put forward a much simpler and more straightforward message. This forum is an occasion to do that, but all the different brain-science fields should bring this message in an integrated way and not everyone in his own corner. And the message is clear. We have to act now because in 2030 it is estimated that 50% of the population will be suffering from a brain disease. The cost of this will be tremendous, so investing in brain research now will be insurance for the future.

You are your brain

SUSAN GREENFIELD: I see three main problems. A brain disease is often perceived as if it is your fault. The second is that you don't necessarily die immediately from a brain disease, unlike with heart disease and cancer. You live with some of the most devastating brain diseases for years and years, slowly towards the end. Thirdly, and most importantly, people are frightened because they don't understand the brain and they know that even for scientists and doctors the brain is the least understood organ. On top of that, they are frightened because brain diseases erode your personality, unlike a cancer or heart disease. A lung cancer might be a horrible disease, but you remain the person you were. With brain diseases you are not the person you were. This scares people and that leads to an ostrich mentality: if you ignore it, it will go away. The more we, as brain scientists, can talk to people about the brain and the more they get the simple message 'you are your brain and the brain is what you are', the better. And we are starting to understand technically how the brain is what you are and that really excites people. That is clearly the thing that engages people most because they feel that the new brain sciences are giving them insights into themselves.

Finally, it is all a matter of empowerment. With the message of genetic technologies and the concept of genetic determinism, people feel that it is out of their hands, that they are helpless. If you can tell them 'the more you exercise your brain, the better it is for you, even if you have a faulty gene, the environment has a major impact', than people feel empowered. That is an important message.

The social component

STEVEN ROSE: I am going to break ranks with my neuroscientific colleagues. I don't think that neuroscience is underfunded, certainly not in Britain and the US. It might be in certain countries in Europe, I agree. The real problem is that there is an awful lot of 'snake oil' being sold, and promises are made, which we are not able to meet at any point. Until now the basic neurosciences had very little to do with solving the problems that our psychiatric and neurological colleagues have been talking about. The situation is changing because of the techniques which are being developed now. But particularly in the neuropsychiatric area we have other problems. It is not true that the pharmaceutical industry is underfunded in its research.

Very often we don't need neuroscience to solve problems at all. In the case of addiction and drug use, for example, we don't need neuroscience to understand what we have to do about these problems. We need a great deal more social science, we need proper legislation. Let me give another example. According to the WHO, by fifteen years from now, depression will be a major epidemiological problem worldwide. If we don't understand the social context in which depression is taking place, than all the neuroscience in the world is not going to get us any further in solving that problem. People are not stupid, they understand that.

JAN STAMAN: But are you saying to the public: 'don't wait for us, neuroscientists? We have nothing to offer?'

STEVEN ROSE: No not at all. For solving certain clear-cut diseases, neuroscientists have a lot to offer. Take Alzheimer's disease. It is my opinion that within five or ten years from now, there may be very important new developments which will be incredibly helpful in treating patients and preventing the disease. The same for Parkinson's and other neurological disorders. But not for the psychiatric disorders.

PAOLO LUCIO MORSELLI: I disagree with the last remark. Today, with proper use of the available drugs and cognitive behavioural therapy, we can bring 80% of patients with depression to a normal life. The public does not know this. As a result, fewer than 20% of depressed patients receive proper treatment. I agree with previous speakers that psychiatry is being discriminated against also at the level of funding of psychiatric research. Despite the fact that in each family there are psychiatric patients present, it is extremely hard to get psychiatric research funded in Europe.

STEVEN ROSE: Professor Morselli did not answer my point why there is such an increase in the diagnosis of depression across the world. We do not understand the epidemiology of depression and we have to begin there.

A gap between disciplines

GRAHAM FARMELO: I am finding the current discussion very educational but what we, as members of the ECD consortium, have to do is to engage the public on the key issues in brain science that we and you think are important now and in the next ten years. Frankly, and I don't want to be insulting, funding of research seems to me to be secondary. All scientists want more funding, and I don't blame them. But we have to go to the lay members of the public and say: 'these are the issues as regards the way your life could potentially change as a result of the work which is being done in neuroscience'. We have to compile that list of things.

GERRIT RAUWS: There are two things that have struck me during this discussion. The separation of the different disciplines is admitted by everyone as a serious barrier, not only to getting more money but also to even solving the most fundamental problems. Apparently, not all brain scientists speak a common language. We even have to question whether the concept of brain science really exists at the moment. Secondly, I observe an enormous gap between fundamental, basic science and its medical applications.

JAN STAMAN: Does that gap mean that the medical neuroscientists, the doctors who are actually treating patients, do not really know what is going on in basic neuroscience? On top of that, you are allowed to take risks with your treatments, because there are no alternatives for the patients. If the patient dies (because of the treatment), well bad luck, he was dying anyway. Is it like that?

MARK VAN BUCHEM: If you start with the diagnostic clinicians, and I am one of them, than it is clearly not like that. Simply because our diagnostic procedures, mainly imaging and scanning, are non-invasive and there is no risk for the patient. That is why these techniques are not only widely used in diagnosing patients, but also in fundamental scientific studies. The ethical issues in terms of consent and risks are not serious.

JAN WILLEM BERKELBACH VAN DER SPRENKEL: I agree with the statement that there is a gap between basic neuroscientists and clinical neuroscientists. I feel that gap, even today. There is a difference in approach. We are treating diseased brains and not healthy brains. With a diseased brain you are indeed allowed to take more risks, but they are calculated risks. If you don't do anything, you know that things will get much worse. But we are evolving rapidly. Twenty years ago, many more patients were dying from neurosurgical procedures. But now hardly anyone dies during the procedure itself anymore, so a lot has changed.

A new cosmology

ANDREAS ROEPSTORFF: I would like to come back to the theme of the discussion, namely possible developments in brain science. In brain science, people are now stating officially what has been an underlying trend for a long time, namely, that 'you are your brain'. As a framework, this is going to be a major way in which people will understand themselves. That is really why problems and issues relating to brain science are becoming much more important to society than some time ago. As long as brain science had to do with understanding how neurons work, with which genes were transcribed and with treating diseases that might be difficult to treat, the implications were really not that big. But once ideas from brain science come to formulate basic tools for what it is to be a human being, then the situation changes. This is what is going on for the moment. Brain science is now making cosmology. Brain science is the new cosmology, and that opens up a whole new set of questions. Brain science is about more than treating patients.

BERIT FABER: I agree with Andreas Roepstorff. Susan Greenfield's statement 'We are our brain' is the promise and the peril in the discussion of brain science with the public. This touches upon questions of the new religion. In the old days, it was believed that the soul was located in the heart. Now the situation is different. We, as the public, are afraid that doctors and scientists tamper with the brain because the 'new religion' says that our soul is seated in the brain.

Session 3

| Views on developments in brain science from a societal perspective |

Introduction

The second day of the workshop was devoted to the listing and mapping of the societal and ethical issues connected with the brain sciences. It was a very intensive exercise of imagination and *conciseness* for all the participants.

Twelve experts and stakeholders had been invited to prepare a short statement about what they considered to be the most pressing issues for citizens in Europe to debate. After each statement, the moderating science journalists recapitulated the issues with the speaker and charted them, with the help of the audience, in categories defined beforehand. A notice board was designated for each of the categories and the moderators wrote the separate issues on the different notice boards. The predefined categories were - in random order: 'limits of normalcy', 'social justice', 'privacy', 'personal freedom', 'personal identity and consciousness', 'commercial and military interests', 'safety and ethics of research', 'responsibility and free will', and 'reductionism'. The exact wording of these categories had to be changed during the process, and new categories were established when a statement was made that did not fall under one of the existing categories.

Such a process had no pretensions to be scientifically exhaustive but rather to be a dynamic exercise and a confrontation of opinions and feelings from experts with very different - and complementary - backgrounds. In other words, 'brain-storming'. Speakers could not have the time to list all the issues they deemed important; they had to make a selection and to trust that their colleagues would pick up other pressing themes. The listing of the issues, made collectively and extemporaneously, could naturally not be exhaustive either, and each item could have been further discussed for hours, although the moderators did keep up a very strict pace.

At the end of this two-and-a-half hour mapping exercise, one wall of the meeting room consisted of ten notice boards (each representing one group of related issues) which were covered with close to fifty different societal and ethical issues related to the brain sciences. After this mapping exercise, a voting session took place to decide which of the categories or issues would be further discussed in depth. The three topics that attracted the most votes (limits of normalcy, reductionism, and responsibility and free will) were put to a debate immediately after the vote. The transcript of these three discussions appears in session 4 of these proceedings.

For a more structured and synthetic view of the socio-ethical issues involved, the reader can refer to part II of the report, which is a synthesis and a further analysis of the results of this workshop, combined with a review of the international literature on the subject.

Statement by
Professor Steven Rose

Professor Steven Rose is a neurobiologist and Director of the Brain and Behaviour Research Group of the Open University (United Kingdom).

My starting point will be a statement that was made yesterday: 'You are your brain'. And, with respect, I want to say: 'Hang on for a moment. You are more than your brain'. You are more than your brain for the important reason that organisms with a brain, and especially humans, are exquisitely bio-social creatures. We can only truly understand the nature of being human by acknowledging that organisms are created evolutionarily and developmentally. If we do not take this developmental perspective into account, which creates human persons from the fusion of egg and sperm and leads to the 100 billion neurons in the brain and the 100 trillion connections between the neurons, we will not understand anything.

Personally I would like to transcend those old dichotomies of 'mind and brain' and 'nature and nurture'. Nonetheless, one of the issues in the neurosciences which concerns me extensively at the moment is precisely the reductionism that goes along with the arguments which reduces mind to brain and then reduces brain to molecules or to the inevitable products of genes. This reductionism produces a set of issues which is tremendously important for this meeting and the subsequent meetings to consider. It might, for example, lead us to an erosion of human agency and responsibility. I want to argue for the restoration of our concept of responsibility. Being bio-social organisms - that is, having a mind that is wider than the brain - enables responsibility. It doesn't reduce responsibility to the notion of molecules.

The fact that many of the reductionist statements are wrong or poor science, as I may argue, does not necessarily mean that they don't have a technological or social impact, both for the way that we think of ourselves as people and even more for the way that we develop technologies in order to approach both ourselves and our fellow citizens. That is: the technologies may work, even if they are based on bad models.

The general issue of concern to me is that neuroscience and neurogenetics are increasing the power of a tendency to what I would describe as the medicalisation of social problems. If you look at the current diagnostic and statistical US psychiatric manual, the DSM IV, you will find a range of new disorders listed, like: conduct disorder, oppositional defiance disorder, attention deficit hyperactivity disorder, age-associated cognitive impairment. That is to say: a listing of a whole range of conditions which are redefining what it is to be a normal human being. A range of human activities are declared 'abnormal' and medicalised, and these activities consequently need to be located somewhere in the body.

One of the consequences of this is the development of technologies for behaviour control. At the moment these technologies are primarily coming out of the pharmaceutical industry, though this is not necessarily the pharmaceutical industry's fault. The one that concerns me enormously is the invention of a category called Attention Deficit and Hyperactivity Disorder (ADHD), an originally American diagnosis that did not exist in Europe until about ten years ago, but which essentially results in the medicalisation of increasing numbers of young children using the drug called Ritalin. Ritalin prescriptions have gone up in Britain (as elsewhere in Europe), from 2 000 a year in 1990 to something like 160 000 a year the last time I looked at it. The increasing tendency to assume that the right way to treat a problem is to find a drug to manipulate the brain, seems to me to be a tremendously important and tremendously worrying social issue.

"The tendency to assume that the right way to treat a problem is to find a drug to manipulate the brain leads to an erosion of human agency and responsibility."

What goes along with this is the development of cognitive enhancers (something I work on in the lab myself): drugs originally developed in the context of Alzheimer's disease. We do need to consider the ethical legal and social constraints under which these types of drugs are used and are available.

What also concerns me is the thrust towards pre-emptive genetic or brain image screening, the possibility of using brain imaging in order to, for example, identify potential psychopaths. I refer to the American tendency at the moment to use the technique of brain fingerprinting [→ glossary] which, they claim, will identify potential terrorists. We must look at the implications of this.

Discussion

IDA ANDERSON: I would like to ask Professor Rose why reductionism in the brain sciences is so strong during this historic period.

STEVEN ROSE: I can see the following reasons: the increasing power of the technologies being developed, especially of the genetic technologies and the power-charged discourse that comes with genetics; and the failure of the vision which social democracies and socialist countries shared in the 1960's and 1970's, namely, that is was possible to use social engineering to produce a more beautiful world. If we can no longer produce a more beautiful world by changing the one around us, not even in wealthy western Europe, then the choices are to say: 'OK, we go into individual solutions and look inside the individual and personal life-style choices', which easily leads to genetic explanations of problems; or you go into fundamentalist religion. The choice is yours.

MAPPING

ISSUE ►	HEADING
mind = brain = molecules	reductionism
erosion of human agency and responsibility	reductionism
medicalisation of social problems	limits of normalcy
development of technologies for behavioural control	personal freedom
cognitive-enhancing drugs	limits of normalcy
brain fingerprinting	privacy and personal freedom
brain imaging leading to pre-emptive diagnostics	reductionism and limits of normalcy

Statement by Dr Zoltan Nemes

Dr Zoltan Nemes is a psychiatrist and neuroscientist at the University of Debrecen in Hungary.

In my research, I deal with the biochemistry of Alzheimer's disease which is a sort of accelerated brain ageing. Ageing is a natural phenomenon which human civilisation has recently medicalised. Brain science has made significant advances in the understanding of Alzheimer's disease and other neurodegenerative conditions. An etiological treatment of Alzheimer's seems to be within reach; these therapies might become available in my lifetime.

"Possibilities to arrest cognitive (aging related) decline will create a gerontocratic and less competitive society."

This means that these treatments can be used for mitigating or arresting age-related cognitive decline. We will not need to retire at all, we can die directly at our office desks, just as kings and popes do. The longevity of senile brains will make societies gerontocratic and less competitive, because the old will remain in competition with the younger generations. Ageing or senility is associated with conservatism, rigidity, protection of the social establishment at the expense of flexibility. That largely contributed to the death of the Soviet empire.

I foresee that the younger generations will be very frustrated by the fact that they can access property, social influence and political power only when the generation of gerontocrats, my generation, will die out. This generation will, by estimation, be at least 70 years old.

Just as the effective medical treatment of infectious diseases largely contributed to the horrible overpopulation and resulting miserable life conditions for several billion people world-wide, this issue, which is very appealing and beneficial to the individual, will result in harmful consequences to society and will badly hit the most developed countries.

Discussion

JES OLESEN: I think you didn't mention the fact that now it is possible to engineer genetically lower animals to have a much longer life. That just adds to what you have been saying. So the problem may be even much bigger, because methods for prolonging life might become available.

ZOLTAN NEMES: Genetically manipulating humans will be under a legal ban during the following couple of decades. Taking a medicine against ageing is not your concern until you start ageing. That's why the majority of people can be convinced that basically it is no use living for 200 years. But, let's say, in the final two decades of your life, you start worrying about how much you missed. Then there you are going to the physician and ask for medication against ageing. I don't think prolonging life is in itself a problem. The problem is the turnover of generations which is the result of it.

MAPPING

ISSUE	► HEADING
frustration of the younger generation	**social justice**
ageing versus employment	**social justice**

Statement by Dr Berit Faber

Dr Berit Faber is a lawyer and Executive Director of The Danish Council of Ethics.

We are actually having the same discussion as the one among young people in Denmark. Last year, the Danish Council of Ethics invited 17 young people from the 9th and 10th grade of Danish schools to discuss the perfect person. They made a statement on the bio-technological transformation of people. I would like to say that you don't need to be so pessimistic: I think we can and will be able to engage the public in the debate.

They discussed the power of beauty, cosmetic surgery and the question of enhancement and cyber technology. From this discussion I can tell you that there was a general agreement that we should really care about the issues linked to enhancement. They said: 'When parents choose a child's school, after-school clubs and so on, a dissatisfied child can change the parents' choices as soon as they come of age. We can't do that with genetics. They said: 'You can rebel against your parents, but you can't rebel against your genes'. The exciting thing about being a parent is that you are supposed to teach your child a lot. Knowing it all from the outset seems to be unnatural and boring.

So I think we have pointers that there is wisdom available among lay people and young persons. They should be trained in decision-making with regard to ethical issues so that they will be able to discuss these issues. They will be able to bring something into your circles.

Talking about circles: here we have the inner circle (of experts) and the outer circle (of participants). I would rather like to have us talking in spirals. We cannot just go a full circle and think: now we have done it. We get feedback from all kinds of places where discussion is going on.

"Important concepts are interdisciplinarity, learning, dialogue, timing (ethics just in time.)"

Yesterday we were told some very important things about techniques. This is, of course, important but we should remember also to discuss the uses of techniques and this is where we are today. I want to point out the concept of interdisciplinarity. We are talking to each other within different lines of science. And I think that's very important. I would hope that, talking about future technologies, we will take into consideration enhancement problematics. When we talk about treatment, the issues of privacy and social justice are very important. I think we should look carefully at our current security measures relating to research procedures and informed consent: are they suitable for these issues? We should strive for a 'just-in-time' ethics - timing is very important.

Discussion

KARIN RONDIA: You mentioned that we should be careful about privacy and social justice in treatment. Do you have specific issues in mind?

BERIT FABER: I can't help talking a bit about process now. I think it is very important that we do this, but I also think that it is important that we talk about how we do it. I was very inspired yesterday when hearing the debate among experts, because what I gather is that what brain scientists are studying is to a very great extent 'processes', not 'things'. I would very much like you to take this approach to the field of creating a public debate. Because if you want to enhance social responsibility, if you want to bring this forward, I think we need to focus on processes, on how we do this.

This is why I would like us to be clear about how we discuss issues like the social implications of enhancement. We might discuss these issues in a different way from the one relating to safety and research. When we talk about issues, we should try to find out if these are new ethical issues or whether they are covered by 'old' ethical questions.

MAPPING

ISSUE	► HEADING
security measures and informed consent	**safety and ethics of research**
just-in-time ethics	**safety and ethics of research**
interdisciplinarity	**safety and ethics of research**
enhancing drugs	**social justice**

A new category called 'Dialogue and information exchange' was created to receive the issues relating to 'methodology of dialogue'.

Statement by Dr Ian Ragan

Dr Ian Ragan is Executive Director of Neuroscience Research at Eli Lilly in the United Kingdom.

I am going to give a perspective from the industry. This is a grand opportunity since I am the only one here.

My first issue has been addressed several times already: It is the increasing prevalence of psychiatric and neurological diseases. Granted, there is a debate on whether the increase in psychiatric diseases is due to the medicalisation of daily life or not. But that the increase in neurological disease such as Alzheimer's is due to the ageing population is obvious. I think the issue then is: social justice.

There has been a lot of talk about future new treatments. Treatments for neurological disease will come along for several reasons. There is clearly a major challenge for fundamental research to understand the causes of brain disease. In trying to come forward with treatments, rightly or wrongly we do have to use a reductionist approach. The genetic basis of disease is very poorly understood. Brain diseases are polygenic. Very few studies on human genetics are yet leading to experimental therapies in the clinic and, as we discussed yesterday, there are no true animal models of brain diseases.

So, we failed to translate fundamental discoveries into effective therapies. The cost of bringing a new drug to the market at the moment is about 1 billion dollars and it takes ten to fifteen years. The fault is not in the technology. The fault is in the model that we use for the development of drugs. The Federal Drug Authority (FDA), for example, is just beginning to recognise that medical product development has failed to keep pace with scientific innovation. We need a more effective process for bringing medicines to patients. It would take a concerted effort by government, academia, researchers and the private sector to bring this about. The current model does not work.

This, I think, has led us to a failure to match the expectations of the general public, government and taxpayers as to what pharmaceutical research can deliver. Unrealistic expectations have been exacerbated by governments demanding a short-term return on investments in order to justify spending taxpayers' money - and this is why it is a societal issue as regards research. There has been exaggeration by scientists who need to get grants to do their work and, of course, there has been misinformation in the media.

I am very concerned at the use of therapies to enhance normal behaviour. My job and that of my industry is to produce drugs that are more safe and more effective than the previous generation. The safer they are and the more effective, the more likely they are going to be used recreationally, to enhance normal behaviour. That's a risk. I can't stop improving drugs because that is clearly a good thing, but it will raise questions for society in the future which are now similar to those relating to performance-enhancing drugs in sport. No one ever took a cancer drug because they thought it would give them a good Friday night out.

Two things have not yet been mentioned. The first: we live in a society which is increasingly risk-averse and litigious. It is quite clear in the industry that there are areas of research which we do not undertake because of the increasing emphasis on safety by governments and regulatory authorities - and perhaps in response to views of the public. This is stopping effective research. Society needs to re-appreciate what risk-to-benefit really means.

"We live in a society that is increasingly risk averse and litigious. That is stopping effective research. Society needs to perhaps re-appreciate what risk-to-benefit really means."

The second is the low esteem of real science. I don't quite know the reasons for this. I guess it has something to do with self-diagnosis; one has access to vast amounts of misinformation from the Internet. It is more prevalent in brain disease than in other areas. The ability to do research is threatened by animal rights extremists, by people who do not have a constructive and informed debate on the use of stem cells. The pharmaceutical industry is perceived as being greedy, secretive and defensive and, at times, we very probably have been. We are very rarely credited with the discovery of drugs. An interesting statistic for you may be that of all the drugs that were registered and approved in the US during the 90's - they were 280 or something - 93 percent of them were discovered and developed by the pharmaceutical industry.

Discussion

JÖRG NAUMANN: I would like to come back to the high expectations that have been raised in the public and in the media regarding the effects of brain research. What is the reason for these high expectations?

IAN RAGAN: It's inevitable. Many of the things that happen in brain research are so extraordinary. They make very good material for supplements to newspapers on Sunday and that kind of thing. Every time there is some major breakthrough, you hear about it on the radio. But when people ask: 'When is it going to produce a treatment?' scientists start to come back from that, because of the sheer time and effort that are required to translate discoveries into treatment.

A fundamental discovery normally does not even lead to treatment within half a lifetime. But we don't like to say that. Not as a researcher, because it might block your grant and not as a government, because you want to explain to your voters why you are spending money on biomedical research. Everyone is in a conspiracy. Drug companies need to sell their ideas to Wall Street. These Wall Street people know nothing about science whatsoever. Yet we are entirely bound by their views and we have a responsibility to our shareholders. It's neither good nor bad. It's the way things are. It's hard to say how you can get out of that.

MAPPING

ISSUE	► HEADING
increasing prevalence of neurological and psychiatric diseases	social justice
excessively high expectations regarding science and excessively bad reputation of industry	dialogue
drugs to enhance normal behaviour + recreational use of drugs	limits of normalcy
risk and benefit awareness	safety and ethics of research and dialogue
A new category, called 'innovation', was created for the issue 'excessively high expectations'	

Statement by
Professor Paolo Lucio Morselli

Professor Paolo Lucio Morselli is a psychiatrist and currently the Secretary General of the Global Alliance of Mental Illness Advocacy Networks, GAMIAN Europe.

What I want to say is very simple. It is what 72 patient associations think; I'm talking on their behalf. Today culture and society have an important role to play in the way information is communicated. We live in a society where the anti-science attitude often prevails, and brain disorders are still strongly stigmatised and frequently rejected on the basis of prejudice and misinformation.

Two years ago, we did a survey of 3 400 brain-patients. We asked them: what are your wishes? 90 percent answered: 'better information and more research on the brain'. But in actual fact we are overwhelmed by a mass of data, and there is a poor understanding of their meaning. We work on misinformation and wrong hopes. Unfortunately, the mass media contribute very much to this misinformation, together with investigators who release information in a way that cannot be understood.

"We should not medicalize social issues, but we should also not socialize medical problems."

We think that a new dialogue must be initiated. We must take care not to medicalise social issues; but we must also take care not to socialise medical issues, as happened in Italy where the life of thousands and thousands of families has been totally destroyed by law 180,which declared certain psychiatric afflictions to be not a medical but a social disease.

We think that patient advocacy groups may help in filling the gap. But they need help in increasing and improving awareness of brain science and of the meaning of brain science. There is evidence that a mature, well informed and well educated patient can greatly reduce the present burden of the disorder. A well informed patient helps to reduce care costs and so could help in saving a lot of money. But many physicians do not seem to want informed patients.

Because of the poor state of information, politicians, policy makers, administrators and lay persons have a very low awareness of many of the problems linked to brain diseases. They have very little information on recent discoveries. This leads to long delays for consent to apply these new discoveries. In the opinion of Gamian Europe, more timely and more apt information on new discoveries is therefore absolutely necessary.

Discussion

BERNARD REBER: You speak about an anti-science attitude. Can you give us one example of that?

PAOLO LUCIO MORSELLI: One or two days ago the French daily newspaper Libération published an article which totally destroyed the outcomes of an objective double blind trial showing that cognitive behavioural therapy is much more effective for depression and anxiety than psychoanalysis. And in Le Monde's issue of yesterday a former psychiatrist turned journalist was telling us that brain imaging is all fantasy, that we would never get a spider out of a hole by doing brain imagery. This is the anti-science attitude.

KARIN RONDIA: Isn't it the role of the lay press to bring to the public the controversies that prevail among scientists?

PAOLO LUCIO MORSELLI: Yes, but there are many ways of reporting. There are good and bad scientific journalists. Sometimes they report in a triumphant way which generates wrong hopes. Or they adopt a totally negative attitude. Last January, the FDA approved the use of antidepressants for children. This means saving the lives of thousands and thousands of adolescents who won't get into heavy drugs. Nonetheless, many European newspapers were strongly against it.

KARIN RONDIA: Isn't this a societal issue?

STEVEN ROSE: It seems to me that you are arguing for a monolithic statement: there is scientific truth and it is the job of the journalist to report on it. These are issues of very serious debate and I don't think that it is anti-scientific to be critical of the use of drugs in young children. It is an issue of debate and of concern.

PAOLO LUCIO MORSELLI: This is an emotional issue. If a child is sick, you have to treat it. If you have pneumonia, you get antibiotics.

STEVEN ROSE: If you want to suggest that there is a fundamental logical distinction between a young adolescent taking XTC and a young adolescent taking Prozac, than I would be inclined to be a little uncertain about this logical categorisation. The issue is: these are genuine debates, they are not anti-scientific in themselves. And as for the media: we should not shoot the messenger. Sometimes scientists themselves provide these triumphant press releases which are turned into media headlines. I think the fault is sometimes ours as well.

PAOLO LUCIO MORSELLI: I would like you to meet some of the parents of these children.

PAOLO VALERIO: When comparing the care of pneumonia and that of depression, Professor Morselli probably adopts an ideological position, not considering the still very alive scientific debate about the use of psychodrugs by adolescents. Many researchers today believe these psychodrugs are really contra-indicated in childhood and adolescence. If you give psychodrugs in the same way you treat pneumonia the risk is that the psychiatrist, instead of taking care of the mental disorder, makes use of a tranquilliser in order to avoid a psychological consequence of the trouble he is supposed to treat. It seems very important, first of all, to try to understand the patient's psychodynamic features. The only condition where a psychodrug is pertinent is if the psychotherapist notices extreme anxiety in the patient and believes that this cannot be solved with the traditional psychotherapeutic means in time to prevent further mental disorders. As you can see, when you make a therapeutical choice ideology gives you a perspective on how to cure people. According to the psychodynamic perspective, the symptom is a mental solution the patient creates in order to avoid more serious conflicts and psychic pain. Conversely, according to the organic perspective, the symptom is only the sign of an ill body which can be treated with the corresponding drug.

Another discussion followed because some participants wanted to emphasise the need for dialogue and information, but the moderators objected that this was not the meaning of this workshop.
The need for dialogue and information exchange was indeed the very reason for the whole ECD project, but the aim of the Amsterdam workshop was precisely to focus upon what this dialogue and information exchange should be about. Methodological aspects of the dialogue should be left to methodological experts and would be the subject of other discussions.

Nevertheless, it was clearly pointed out that there was a mismatch between expectations that are created amongst the general public and the actual state of affairs. Experts should keep in mind that dialogue is not a one-way street and that they are also going to learn something. It should also be made clear to the general public that some expectations are not realistic.

MAPPING

ISSUE	▸ HEADING
anti-science attitude which is the consequence of bad information and bad dialogue	dialogue and information exchange
need for better information for the public	dialogue and information exchange
need for informed patients	dialogue and information exchange
use of psychotropic drugs in children	limits of normalcy
ideology	reductionism

Statement by Mr Alistair Newton

Mr Alistair Newton is Treasurer of the European Federation of Neurological Associations, EFNA.

I'd like briefly to introduce EFNA, because it is a slightly unusual concept. All of our members are already pan-European federations in their own right. We have members such as Alzheimer Europe, European Parkinson's Disease Association, European Dystonia Federation, Multiple Sclerosis Platform, International Bureau for Epilepsy, Gamian Europe. We work together very successfully, not only with other patient groups, but particularly in collaboration with medicine, science, commerce, media, industry and pharmacology.

"The lack of public awareness leads to stigmatization and discrimination of people with neurological disorders."

Our aim is to promote a wider understanding of how brain disorders impact the lives of people that live with them. There is an immense and escalating economic and social burden of brain disorders. We talked about it yesterday and today. We feel that society has failed to create compatible opportunities for people living with these disorders so that they can actually continue to contribute to their community. Perhaps even more importantly, costs are escalating at a tremendous rate. Professor Olesen said yesterday that the burden of brain disorders in society in health care costs is running over 30 percent.

As for the key issues: first of all there is the great number of people living with brain disorders and a lack of public awareness which leads to stigma and discrimination. There is certainly a need for more specific medical training, especially at general practitioner level. There is a need for better treatments which take into account 'quality of life' issues. There is a need for greater involvement of the patients in the management of their conditions. We've heard a great deal about the 'informed patient', which is a very valuable concept, but I think we have now moved beyond that. We are now with the 'involved patient'. Patients need to be involved in the management of their condition.

Yesterday Gerrit Rauws in his introduction spoke about citizens' responsibilities to inform the researchers of their hopes and fears. Yes, it is their responsibility. However, the researchers' responsibility is to really listen. That's partnership.

Discussion

IAN RAGAN: I have a point which refers to the last two presentations about the involved patient. My industry is very keen to listen to patients. But there are people out there who think that the information from drug companies to patients should be filtered because of our perceived narrow commercial interests. If you are going to have informed patients and they are going to have access to plenty of the snake-oil stuff that Steven Rose refers to, I think they should have equal access to information from the pharmaceutical industry and be allowed to make up their own minds.

"We are now beyond the informed patient. The patient needs to be involved."

ALISTAIR NEWTON: I entirely agree. Not only do we have to find greater educational opportunities for the medical professional, we also have to find them for the patients.

UNIDENTIFIED VOICE: I think these last statements touch upon a very interesting area: the expansion of the Internet. The problem of quality control of the information on the Internet becomes exceedingly important. I don't know how we should handle it, but it is particularly important in this area.

JES OLESEN: I am just wondering if we have enough respect for the ability of patients to understand and make decisions on their own behalf or whether we are still living in a society that is paternalising patients. As a doctor, I am afraid I have to say 'yes' to the latter. Many, many doctors are still regarding patients as children who cannot really tolerate the full truth and who cannot really make a decision on their own. I am very much against this attitude. I know from my work in Europe that there is a big geographical difference. In the Scandinavian countries and maybe in Holland and the UK, in this respect we are a little bit ahead. When I was a young doctor, you never told a patient the truth about a cancer or anything. Now, at least in Denmark, we have come to the point where they must have full information. I don't know where to fit this in.

MAPPING

ISSUE	► HEADING
stigma and discrimination of brain patients	**social justice**
informed patient was extended to informed and involved patient	**dialogue and information exchange**
researchers should really listen This issue was joined to the already existing need for better information for the public to give a new wording: public understanding of science & scientists' understanding of the public	**dialogue and information exchange**
autonomy of the patient	**responsibility and free will**
access to information	**personal freedom**
quality control of information	**dialogue and information exchange**

Statement by Dr Wrye Sententia

Dr Wrye Sententia is Co-Director of the Center for Cognitive Liberty & Ethics (USA)

"Some of our most fundamental human freedoms and democratic rights are shaken up by developments in brain science."

The Center for Cognitive Liberty and Ethics has a mission to protect freedom of thought in the 21st century. We frame it that way because this is an age of accelerating technologies. William Gibson, the cyber punk author, said: 'the future is already here, it's just unevenly distributed'. I think that speaks to the state of technology in relation to the sciences and the public perception of it.

The world is currently experiencing an acceleration in ways to monitor, manipulate and also augment cognitive function. The contribution that I and my Center bring would be to look at how some of our most fundamental human freedoms and democratic rights are being shaken up by discoveries in brain science - I allude to our perception of selfhood, and our ability to have agency.

I want to begin by speaking about memory pharmaceuticals, drugs that are being used to improve attention, alertness and recall. The financial firm A.G. Edwards has estimated that the global market for memory pharmaceuticals is already over 10 billion dollars. Eric Kandel, a Nobel prize winning memory researcher at Columbia University, has said that memory drugs will probably be safe and effectively available within five years. Now Kandel heads his own company 'Memory Pharmaceuticals' that just went public on the stock market two weeks ago and was very successful. These companies produce medicines meant for mild cognitive impairment, Alzheimer's, and things of this nature. But that's not where they will be used. Once a medicine is approved and accepted, other people will use it for other purposes.

Consequently, these sorts of memory drugs raise a lot of novel ethical questions. Do you have a right to dim or erase your memories? There is a drug called 'Propranolol' that is currently being used in conjunction with post-traumatic stress, with accident victims or emergency room attendees. It helps people to forget if given within six hours of the event. Should hospitals be giving this as a blanket procedure when people come into the emergency room? There are some instances when we want to retain that memory. If you are the sole witness to a crime, will the government be able to require you not to take a forgetting drug until after you testify? Or could they even feasibly require you to take a memory drug to keep your memory fresh until the trial? These are novel questions.

How many people have heard of brain fingerprinting [→ glossary]? It is basically a 21st century polygraph that uses EEG P300 wave testing to assess whether or not a person has a particular memory or not. The company that is developing this technology is pushing it as a law-enforcement tool, a pre-employment screening device and, of course, an anti terrorist screening device. Another company, The Brighthouse Institute for Thought Sciences, is developing a novel form of marketing research, called neuro-marketing. Just last week there was an article in the New York Times that neuro-marketing is used to find who will be the next president of the US. I get the impression that 'neuro-' is the new tag; for the next years, it is no longer 'e-', it's 'neuro-'.

There are a number of commercial and military applications that I want to signal. And just as memory drugs raise novel questions, these sorts of applications of brain scanning or brain imaging devices also raise questions. Should a company be able to pre-screen its employees? Should you have a right to say 'no'? If you're accused of a crime, will you be able to say 'no' to a brain fingerprint? Already fingerprinting is done liberally when people are arrested, at least in the US.

Hypersonic sound [→ glossary] is another technology. Hypersonic sound is completely silent. It's a technology that uses the human eardrum as a speaker. When you walk into a hypersonic sound path you literally hear voices or whatever is being transmitted inside your head. If you're one metre away, you hear nothing. It's only directly in the path. Do we have a right to auditory integrity? Is it OK for us to hear voices that are not our own? It might be a good idea for McDonald's in Florida to use hypersonic sound in its drive-throughs so that the entire neighbourhood will not hear the attendant saying: 'Would you like fries with that?' That's a good application, I think. However, there are applications that are not so good. The US government again is interested in hypersonic sound as an auditory weapon.
What does this have to do with freedom of thought? The Supreme Court has recognised freedom of thought, as does international law, but they haven't articulated it very well. The United Nations also recognises freedom of thought in their Declaration of Human Rights. They have in article 18 a recognition of freedom of thought, but it is qualified largely in relation to religion. This raises a lot of grey areas. And at the Center for Cognitive Liberty and Ethics we feel that a lot of these developments will best be guided by an application and regulation of the neuro-technologies that respects the individual's fundamental right to freedom of thought. I would say that neuro-technologies are now to freedom of thought what the printing press was to freedom of speech.

"Neurotechnologies are to freedom of thought what the printing press was to the freedom of speech. Freedom of thought must be respected by neurotechnologies."

I invite you to ally with the Center for Cognitive Liberty and Ethics as we fight for freedom of thought. The three issues I want to flag are: privacy, autonomy and choice.

Discussion

AXEL CLEEREMANS: You mentioned a lot of issues that have to do with control of the public: commercial interest, the military, etc. But I think the public shares some responsibility in the sense that once a technique or once cognitive enhancement becomes available, it will be used by the public. Just as cosmetic surgery is used now, even if it isn't strictly necessary. The point is: when these things become available, people will use them. The issue is: should we regulate that or not?

MAPPING

ISSUE	► HEADING
memory-enhancing drugs	**limits of normalcy and commercial interests**
memory-erasing drugs	**responsibility and free will and personal freedom**
brain fingerprinting	**personal freedom and privacy**
hypersonic sound	**commercial and military use**
neuro-marketing	**commercial and military use**

Statement by Professor Flavio Keller

Professor Flavio Keller is Professor of Physiology and Neurophysiology and Head of the Laboratory of Developmental Neuroscience at the Bio-Medico University Campus in Rome (Italy).

I prepared my statement in collaboration with Dr Emilio Mordini, and it deals with the ethical, legal and social issues that are related to the mind-body relationship.

One can safely state that the human body in general and the human brain in particular (and we shouldn't forget that philosophically speaking it is part of the body) appear to be uniquely human. Questions related to this statement are: up to which point can some parts of the body be substituted without overturning personal identity? Should organs from animals be transplanted into humans or not? Is there an ethical necessity to place some limits on the physical manipulation of the body or brain in order to safeguard personal identity?

The mind-body relationship is not simple and constant. It is something that changes from individual to individual and is dynamically evolving in time. Legal issues that can arise from this statement are: can a person be prosecuted for a criminal offence that he or she committed a long time ago? What should the legal and moral responsibility be of people suffering from multiple personality disorders?

"We still know very little about the language of the mind body relationship (MBR)."

We still know very little about the language of the mind-body relationship. I have developed a metaphor that explains this a little better: suppose that there is a Martian who only knows spoken language. He has never heard of music played by instruments. He is given a musical score - with the musical transcription of the notes - that he obviously will interpret in terms of spoken language. So he can spend thousands of years on this musical score without ever guessing what its real meaning is, because he doesn't know that the language of musical instruments is different from the spoken language.

The point is that there may be fundamental, not just technological, limits to influencing brain activity. One can speak of indeterminism, not in the sense that is known from physics but in another sense that is typical of complex dynamic systems. The exchange of signals below the level of consciousness, which might have mass psychological effects, or the uniqueness of subjective experience might not easily be grasped in terms of the brain sciences. Should subjective experience be granted as much reality value as objective assessments - looking into the brain by neuroimaging or whatever? If not, why not? There is known to be a big difference between the subjective experience of pain and the physiological processes of pain. Does it still make sense to differentiate between mental and organic disorders?

The last issue is maybe the most relevant. Psychiatric conditions may be the result of the interaction of a normal human mind with an abnormal neural substrate. This would lead to the conclusion that mentally ill persons have the same dignity as other patients. In fact suffering in general, and mental suffering in particular, commands in the first place respect, even before we think of treating patients. We need this respectful attitude towards mental suffering. Given the evolution of psychiatric research: are current policies adequate or should additional policies to implement this equal dignity be established? Are there current health policies or treatments that implicitly or explicitly deny equal dignity to mentally ill persons?

Discussion

BERNARD REBER: You spoke about ethical limits to manipulation and the search for personal identity. Have you any answer to submit from your professional expertise and field? It's not only an ethical question, is it? Or am I wrong?

FLAVIO KELLER: I am not sure that I get what you mean. As a developmental neurobiologist, I am biased from my professional experience. If you put a new organ, either natural or artificial, into a body that has already grown up, it is an intervention in the adult life. The animal body in general and the human one in particular starts from the very beginning as a unity. The organs - the liver, the heart - exchange information with each other from the very beginning. How will this new entity inserted into the body, after the early phases of development have finished, influence this communication? In order to decide if this procedure is ethically defendable we need to know whether a personality is changed by this or not.

STEVEN ROSE: Are we arguing that there is an ethical or moral distinction between enhancement, for example, by sending your children to a private school, by buying environmental manipulations for a person in order to improve their life chances - and the technological ones of applying a drug of this sort? I am not sure there is a legitimate distinction we can make. I think we need to address that at some point.

WRYE SENTENTIA: This is certainly a big issue in the US. The US President's Council on Bioethics came out with a report last autumn, 'Beyond therapy and the pursuit of happiness'. In my opinion it conflates a number of issues. The slippery slope between what is an enhancement and what is not, and particularly how to distinguish between so-called natural enhancements, like a preparatory test for college, is arbitrary.

MAPPING

ISSUE ►	HEADING
dignity of the diseased person was joined to stigma and discrimination	social justice
responsibility of persons with mental disease	responsibility and free will
indeterminism	reductionism
vocabulary used to talk about the brain and the mind	reductionism and personal identity and consciousness
techniques that may cause the loss of personal identity + animal transplants	personal identity and consciousness
assessment of subjective experience in psychotherapies	social justice (because this could lead to the end of their reimbursement)
manipulation by mass psychology	responsibility and free will
difference between cultural and technological enhancements	social justice

Statement by Professor Jochen Oehler

Professor Jochen Oehler is a neurobiologist on the Medical Faculty of the Technical University of Dresden (Germany).

I want to talk about the development of the individual mind in early life. Brain structure and its function are the direct result of a continuous dialogue between genetic and epigenetic influences. The brain is developed under the influence of information from outside, which means an external input, and on the basis of endogenously expressed genetic traits. Our behaviour, our later ethical position, is a result of that lifelong interaction.

Within the category of genetic influences we should distinguish between phylogenetic and ontogenetic aspects. Both categories influence the structure-building of the brain. The phylogenetic rules have provided programmes, and these programmes develop during early life (in utero and especially in the post-partum phase). So during ontogenetic development, the phylogenetically acquired programmes (the genetic background) must be triggered and tuned by adapted stimuli.

The interaction between the phylogenetic and the ontogenetic programmes is very variable: large deviations are possible. Especially after birth, the process of maturation of several brain structures is under the influence of external information. We know a lot about the developmental processes in the higher brain structures. But amongst our phylogenetic rules, one is essential: an intensive dialogue/interaction between the new-born and its near environment - the mother, of course, and later on the persons that are near to the child. This continuously supported dialogue is a condition for the normal development of the mind. Defined social contacts are much more important for early development than a lot of artificial toys or pictures on TV.

"Media environments created by ourselves are much more dangerous than simple pharmacological manipulations."

We know that development after birth is different in the cortex, in the limbic system, and so on. What we don't know is, how exactly these internal mechanisms are linked to one other and to externally triggered processes. But we have to hint at the fact that there should be an optimal range of external information input. We are actually living in an overflow of information - an overflow as regards quantity and/or quality. The question is: what are the results of individual information processing if the externally triggered input is inadequate in quantity and quality for the ontogenetically time-dependent maturating brain?

I think we should pay attention to the external information because of its influence on brain maturation. The effect of the media is the problem here. I quote Thomas Metzinger, a German philosopher, saying: "I believe that the media environment created by ourselves is much more dangerous than that created by pharmacological manipulations."

Discussion

PETER MOLNAR: We are doing research on what I call 'the first dialogue', until the first six or seven weeks of life, and this interaction is crucial. It's one of the most important aspects of normalcy. Modern young children are often just given toys and television programmes instead of direct personal interaction. This so-called object-mediated love seems to be one of the most dangerous factors for normalcy in early childhood and empathy development. This might be a very interesting potential negative conditioning of future generations. We know that we are shaped by evolution to have very important face-to-face, unmediated connection with our significant others. The sector of the media is a small thing compared with the input in the first weeks of life.

FLAVIO KELLER: I just wanted to point out that the classical definition of man is 'rational animal'. I think this is not enough; man could also be defined as 'social animal'. You see it in conditions like autistic disorder. The core deficit of autistic disorder is a lack of social communication, which is much more disruptive for the life of the person or the family than, for example, mental retardation.

MAPPING

ISSUE	► HEADING
overflow of information/power of the media further changed into power of words	**limits of normalcy and personal freedom**
lack of personal interaction	**limits of normalcy**

Statement by Professor Demosthenes Agrafiotis

Besides being a social scientist at the Hellenic National School of Public Health (Greece), Professor Demosthenes Agrafiotis is also an artist.

In Amsterdam there is the famous painting by Rembrandt, 'The anatomy lesson'. I'd like to speak about 'The brain lesson'.

"It is wrong to speak of brain *and* society. We should speak of brain *in* and *of* society. Science shapes society, but society also shapes science."

I don't think it is pertinent to speak about brain and society. We should speak about 'brain in society', and 'brain of society', because there is a tremendous difficulty about proposing a distinction between brain and society. There is always very complicated interaction and we have to take care of a multiplicity of mutual influences between 'brain' and 'society'.

There are two ways to think: one can assume that society shapes technology or that technology shapes society. Until now, the discussion has been more about technology or science as forces shaping society. But if you think the other way around, you end up with another type of question. So my question here is: What is society? What is contemporary society? Yesterday we heard presentations about neurosciences, brain research and mental health. Society is even more complicated than these fields, but we have not yet discussed it. We should do it, in order to be symmetrical and honest, scientifically speaking. What is culture today? What is the cultural destiny of our societies? I think there is something missing at a methodological level in this meeting because we are not discussing this crucial issue. Can I ask you a question? If we agree that our society is a society of uncertainty, does brain research bring more certainty or more uncertainty? That's the basic question from a social and cultural point of view.

So, what are the cultural dynamics today? In every society there is a tendency to present an approach from one point of view to explain everything. We say to the people: 'you are your brain', 'you are your genes', 'you are the money you earn', 'you are your childhood'. We always try to give one reason, one cause, for everything. But in our post-cultural societies, meanings are not that clear. Until now, we've always had science without knowing everything about the brain, and we have had a brain without science. From my point of view, let us call it 'socio-cultural'; the crucial issue for our societies is the following: is it possible to elaborate an approach that permits us to think, to analyse and to act in the name of both universality and particularity? Both globally and locally? Both for the past-present and the future? In other words, the challenge is to be intelligent but also sensitive to the complexity of reality. In this respect, the critical question could be phrased in the following way: To what degree can brain research contribute to elucidating the above issues? Could it challenge or realise socio-cultural initiatives inspired by the aforementioned perspective? So, our debate has to go beyond the potentialities of brain research and to be centred on the cultural foundations and orientations of contemporary societies. Based on this assumption, then the fundamental problem in our society is: How can all this brain research help us with these basic questions in our societies? And not vice versa.

When we speak of the great revolution in brain research (like the Darwinian one, I suppose), what will be born from it? What other things will have to die? It could be the end of art, for example, because there would be some kind of hyper-knowledge, because the orientation will be towards grasping everything at the level of the brain, thought and mind.

MAPPING

ISSUE	► HEADING
brain in society and brain of society	**reductionism**
hyper-knowledge, no more art, no more mysteries	**reductionism**
brain research should address societal issues and not vice versa	**safety and ethics of research**

Statement by Dr Andreas Roepstorff

Dr Andreas Roepstorff is a medical anthropologist at the University of Aarhus (Denmark).

I would like to raise one issue even though there are a lot of other issues embedded in it. There is currently an ongoing trend towards formulating a neuronal human. Knowledge about the brain seems to become quite rapidly some kind of major metaphor for understanding what it is like to be human. An astonishing hypothesis, stated by Francis Crick: "You are nothing but a pack of neurons", was the first outspoken attempt at formulating something like this. This tallies with the present trend that everything which had a social component to it in the 1970's now has a neuro-component. We have neuro-philosophy, neuro-linguistics, neuro-ethics, neuro-aesthetics coming up like kinds of acceptable research fields.

This development towards the neuronal human is really difficult to stop. That signifies to me that neurons and the brain are moving out of the hands of the brain scientists and into becoming a kind of important metaphor for thinking about what it is like to be human. We saw examples of this yesterday. I am quoting now from memory. Susan Greenfield said something like: 'You are nothing but your brain." Berit Faber said: "The problem is: we don't like you to tamper with the brain because somehow we believe that the soul is in the brain." This really makes the point that the whole notion of the brain is no longer the domain of the brain sciences because it has become an important metaphor for a whole lot of other issues. Peter Molnar very succinctly pointed out: "You should say that you are not only the brain, there is also the body. And this body is very important and significant to others." We could perhaps rephrase this and say: "Well, if you are your brain, then we'll have to understand the brain and include the body and what is significant to others." That suggests that our understanding of the brain is terribly important for what could be included in these kinds of issues.

I think this lends itself to a conclusion: today the brain sciences are not only about treating diseases, they also form important narratives about what it is to be human. This is a new scenario and that is why it is so important now to have this kind of societal discussions about what it is that goes on in the brain-society interface. What does it mean that something becomes neuronal? I think the last two talks were beautiful examples of that.

There is a kind of alternative to a simple naturalisation here. On the one hand we have a trend that says: 'you are your brain'. But at the same time we can see that the brain is very much the others as well. So the brain becomes not only just nature, but rather a meeting place between nature and culture - if you like these distinctions. That came out very beautifully in Jouvent's presentation yesterday, where he stated that it might not be a question of either pharmacology or psychotherapy. Maybe what we are moving towards is that biomedical and cultural techniques are working together in meetings in the brain.

"The brain becomes a meeting place between nature and culture, if you like these distinctions. Saying 'you are your brain' is naturalisation, but at the same time that nature is heavily culturalised. This is a paradox that requires reflection."

We might see a naturalisation of the mind in the brain, but at the same time this type of nature is heavily culturalised. This is a paradox that requires a lot of reflection. That goes back again to Berit Faber's comments that now the brain is just no longer a matter of neurons, but a space to think about what it is like to be human. We see that a lot of discussions that were classically played out in other fields related to personal identity, religion and experiences of agency are now taking place with reference to the brain.

I think it became obvious to most of us yesterday that science and establishing facts is necessarily esoteric. You have some of the best neuroscientists here and even though they did a tremendous job in trying to simplify the kind of things they are doing, immediately when a discussion pops up, it becomes extremely technical. So basically we have to rely on the fact that they do their work properly.

One way you can make facts robust is by making certain that interdisciplinarity is already there at the level of research. Interdisciplinarity means not only different types of neuroscientists, but also philosophers, anthropologists and psychologists are involved. Because that opens up important discussions at a point before the facts are actually established. This ensures that the community that you, the public, has to trust and believe in becomes larger.

Secondly, I think that it is extremely important to acknowledge that science - and at the moment particularly brain science - is not only about establishing facts. It is also about establishing cosmologies, it is about making stories about who we are, what our place in the world is. And so to hold discussions that relate not only to science but to much wider areas.

If you start out from these two positions, namely, that interdisciplinarity is an important part of the fact-establishing process and acknowledging that the brain sciences are not just about establishing facts that should be disseminated but also about establishing a new cosmology, you create the possibility for a public debate.

Discussion

FLAVIO KELLER: This was a clear presentation of the threats of reducing the mind-body relationship to the mind-brain relationship. This kind of reductionism brushes aside the role of emotions. Because emotions are by definition bodily changes: heart, respiration, transpiration, and so on. Nowadays the cognitive role of emotions is very well known. If one acknowledges this reductionism, one should be aware of brushing aside these findings. It's an important issue to be aware of.

MAPPING

ISSUE	► HEADING
brain becomes a metaphor for what it is to be human	reductionism
brain research becomes a cultural narrative and a new cosmology	reductionism
embedding research in society	dialogue and exchange of information
embedding interdisciplinarity in research	safety and ethics of research

Statement by
Dr Bernard Reber

Dr Bernard Reber is a philosopher and researcher at the Centre de Recherche Sens, Ethique et Société at the Centre National de la Recherche Scientific (CNRS) and the University Paris V (France).

My statement is mainly focussed on ethical perspectives, in which I want to avoid any technical determinism or strong social constructivism. Research in brain science is fundamentally altering our conventional views of human nature and our vision of who we are. How do we think, feel and act? How do the brain sciences affect our concepts of simulation, responsibility, and free will? How, and under which conditions, is it possible to enhance our biological substratum?

Some questions, like those of ethical perfectionism, weakness of will, and reductionism, are not new from a moral perspective. Nevertheless, the intensity of these questions increases as the result of new technological possibilities. These possibilities affect the radicalism of the interrogation, especially when the intervention is not social but directly produced on the brain. Therefore, caution as regards current research presentations and expected results is tremendously important. Partial understanding of such a complex phenomenon as the developing brain can lead in the wrong direction. We have a kind of epistemic and semantic duty before speaking of any consequences in a field which is so heterogeneous and so 'phantasmic' as that of the brain sciences. It is important to understand which correlation can precisely be established in this research. After yesterday's session, I see the brain sciences not only as a cool interdisciplinary field. It's a fight, perhaps a possible war, among scientists with new possibilities of assessment. I am thinking of the debate in France concerning psychoanalysis versus the brain sciences and the big controversies about it. (See the parliamentary bill concerning prescription and guidance of psychotherapies: 'prescription et conduite des psychothérapies'. http://www.assemblee-nat.fr/propositions/pion2342),

"Some anthropological assumptions in brain research could lead to a more deterministic general world view. This shift might have negative consequences for the public appreciation of individual freedom, responsibility and choices."

A couple of our listed issues combine classical bioethics and ICT, for example: privacy in the case of the dissemination of data, justice in access to treatment and clear consent of the patient. But new questions arise. Privacy is high on the list of ethical issues raised by the new brain-imaging technology. But for me, the more pressing social issue raised by scanning is the inability to measure mental or neural processing meaningfully. A better public understanding of the capabilities of brain imaging is necessary to prevent over-reliance on this source of information.

There is a new subfield in ethics: neuroethics. Think about the possibility of free will and choice in the case of brain intervention, and you have a framework for the types of ethical questions that neuro-science is likely to confront over the next 25 years. There is a tendency in brain research towards a more deterministic general view. This might have negative consequences for the public appreciation of individual freedom and responsibility. What are the limits of plasticity in the brain? Is it legitimate or not to defend a moral position based on human nature when we know that 'natural' is not the same as 'good'? How robust is our certitude that our interventions in the brain will be better for the state of the world? Especially if these interventions might not be reversible?

MAPPING

ISSUE	► HEADING
semantic duty + use of incomplete information	**safety and ethics of research and dialogue and information exchange**
assessment of different forms of therapies	**social justice**
natural is not the same as normal + deterministic views	**responsibility and free will**
irreversibility	**safety and ethics of research**

Voting session

After the mapping session was closed, all participants were asked to vote for the issues which they felt were the most pertinent or urgent to discuss with the public. Each expert/stakeholder had three votes, and each country participating in the ECD project also had three votes, irrespective of the number of delegates they had in the workshop.
The vote took place in a very informal way: during the break the participants were free to wander about and to stick their 'votes' - in the form of post-its - on the panels of their choice.
In principle, it was possible to vote either for a category or for a specific issue, but it turned out that most of the votes were put on top of the panels, i.e. for the category headlines. Thus it was decided to count the votes globally per category.

The results were:
Limits of normalcy: 17 votes
Reductionism: 16 votes
Responsibility and free will: 16 votes

These three categories of issues came considerably ahead of the others and were consequently the subject of the three discussions that follow in Session 4.

SUMMARIZING MAPPING MATRIX

Medicalisation of social problems (ex: ADHD)	Language (vocabulary)	Autonomy	Promising too much / hyping	Language (vocabulary)
Medicalisation of normalcy	Mind = Brain = Molecules	Manipulation by mass psychology		Techniques that may cause
Brain Imaging leading to pre-emptive diagnostic	Erosion of human agency and responsibility			identity + Animal transplant
Cognitive enhancing drugs	Brain Imaging leading to pre-emptive diagnostic	Memory erasing drugs		
Drugs to enhance normal behaviour + recreational use	Ideology	Responsibility of persons with mental disorders		
	Indeterminism	Irreversibility		
Use of psychotropic drugs in children	Brain becomes metaphor of what it is to be human	Natural is not the same as good + deterministic views		
Power of the words / of the media				
	Brain in Society / Brain of society			
Lack of interpersonal interaction	Hyper knowledge –> no more art, no more mystery			

Security measures & informed consent	Hypersonic sound	Embedding research in society	Brain finger printing	Brain finger printing	Frustration of the young generation
Ethics in time (and loss of personal	Neuro-marketing not later)	Methodology of dialogue	Development of technologies control	for behavioural	Ageing <–> employment
Risk & Benefit awareness					Enhancing drugs
Brain research should address societal issues & not vice versa	Enhancing drugs	Informed and involved patient	Access to information		Stigma / discrimination of brain patients
Embedding interdisciplinarity in research		Public understanding of science & scientists understanding the public	Memory erasing drugs		Increasing prevalence of neurological and psychiatric diseases
			Power of words / of media		
Semantic duty + Use of incomplete information		Too high expectations on science, too bad reputation of industry			Assessment of different therapies
		Risk & Benefit awareness			Relevant difference between social and technological enhancement
		Anti-science attitude			
		Quality control of information			
		Semantic duty + Use of incomplete information			

Session 4

| Discussion sessions |

Plenary discussion of the 'Limits of normalcy'

WITH AN OPENING STATEMENT BY PROFESSOR STEVEN ROSE

MODERATORS MARJAN SLOB AND KARIN RONDIA: Our first discussion will be about the 'limits of normalcy', which include the enhancement questions. It's about the drift in the definition of what is normal and what is not, and also about the use of medicine for people who are healthy but want to enhance their cognitive functions or take it for pleasure. It also includes medicalisation, medical labelling of conditions that are not necessarily pathological. The aim now is to clarify what sort of concepts were brought together this morning. It's still part of the mapping process, but more detailed and in greater depth. During the break, we asked Professor Rose to make the opening statement.

STEVEN ROSE: It seems to me that there are both theoretical and practical issues which surround the general discussion of the limits of normalcy. The first question could be: how much do we socially accept the range of variation of human behaviour? There is a tendency, which is more marked in the US than in Europe, to actually assume that if you divert by more than 2 standard deviations from the norm, you are abnormal in particular sorts of ways.
But within the concept of normalcy, there is not only this statistical concept but also a normative one, that is: how people ought to behave. The question then becomes one of the extent to which people should be adjusted towards the norm if they are diverging from it. If we don't accept diversity of our children in school, then a child who 'misbehaves'

must be adjusted back into the school environment, rather than for example considering the relationship between that child and its social or educational environment.

Those are the theoretical issues - I am improvising here because I didn't know you were going to ask this of me. This is where the question of medicalisation comes in. What do you do about a particular behaviour if you define it as abnormal? This is where the issue of ADHD and Ritalin becomes extremely interesting and important. You could argue that a child misbehaves in school because of a problem in the relationship between the child and the school or between the child and the parents or teachers or also the lack of books, the social context of education, etc. Medicalising this situation puts the responsibility for treatment and correction of the problem back on the child, and more precisely back on the brain of the child. Now I don't want to say that Ritalin does not work. If you have a toothache, you can treat the toothache by taking aspirin because it relieves the pain. It doesn't cure the problem, it masks the problem. My concern is that some uses of drugs in this sort of way would be masking a problem. There can be a reason for masking a problem while you get the proper treatment. But there is no reason for going on taking aspirin and not dealing with the teeth. If a child's misbehaviour at school is defined as a problem, we need to look at the causes instead of simply masking it. So that's one side of the discussion.

The second side of the discussion concerns the issues of enhancement. A clear example is the case of cognitive enhancers. The treatments we are developing in the laboratories for Alzheimer's disease are going to bring clear benefits in terms of treatment of the disease itself, although not entirely, because there are issues that gerontologists will raise about whether one wants to enhance the patients cognitively rather than allow a certain progression of the disease. Maybe we should discuss that in the context of what we were talking about earlier this morning, i.e. ageing and the nature of ageing etc.

But these drugs also bring us to the enhancement issue in the context of what Ian Ragan called the recreational use of drugs or the use of drugs in competitive environments and such things. Here we are moving into a very complicated situation. I think we need to look very carefully at the legal and social implications that come with that. The problem we've got there is regulation: you can buy everything from the Internet, snake oil and effective drugs as well. Whether it is possible to regulate this, I don't know. If you look at the extent to which the athletics councils have difficulty in regulating the use of drugs in athletics, then you begin to get a picture of the problem.

What is very clear to me is what's happening in the neurosciences and neuropharmacology. Someone mentioned yesterday Aldous Huxley's famous 'Brave New World', where there is a universal drug called 'Soma' which everyone takes in order to keep happy all the time. What we are moving towards is a situation where we will be in a pharmacologically enhanced or available universe in which everybody may have their own personally tailored Soma. We need to actually think very carefully, I suspect, about the implications of that. Normalcy will no longer be a natural thing, it will become a biosocial condition.

MARJAN SLOB: I would like to go to Ian Ragan now, because he is the representative of the pharmaceutical industry. Do you have an opportunity of talking about these sorts of issues within your company or is it something you do in the pub?

IAN RAGAN: Something we discuss all the time in the company as well as outside is: what are we trying to achieve? I will make a couple of comments. Let's get back to the ADHD issue. And I should be quite open here and say that my own company has a very strong interest in ADHD. They produced a new treatment for it which is not a psychostimulant. It's a drug called 'Stratera' and it has recently been launched. So I can't be completely unbiased.

We should not throw out the baby with the bath water. The overuse of drugs should not prevent them from being used where there is a real need for them. And I don't think any psychopharmacologist has any pretension that drug treatment is the answer to the problem. We talked about disguising the condition. But if it enables a certain amount of control over the behaviour of the child such that other approaches to the condition can be used, then it is useful. So it is merely an aid to the treatment of the child. That's true for ADHD particularly. It's true for Alzheimer's disease as well. It's a question of helping the condition, it's not the answer to it.

KARIN RONDIA: Isn't this wishful thinking? We know that a drug like Ritalin is not only used to facilitate other therapeutic approaches. Everybody has heard stories of parents put under pressure and of children not allowed in schools unless they take the drug. This is not necessarily directed towards industry; society itself should be questioned.

JOHAN DEN BOER: I think we should not confine ourselves to the history of ADHD, because if we look at the history of psychiatry, there are a tremendous number of other diagnostic entities which are simply based upon consensus. In the DSM IV there are 297 disorders. For instance, the theory of multiple personality disorder was formerly considered to be vague or almost non-existent. Now we can clearly show that there are real biological differences between people with MPD and normal people. The same has been true with patients suffering from social anxiety disorders. Sociophobia has been seen as some kind of hypertrophic shyness, and now some pharmaceutical companies are earning millions and millions of dollars selling SSRIs to patients who suffer from this disorder. It's a wonder that we are all sitting here being psychologically normal.

There have been discussions about that, at least in the Netherlands. They said: isn't it true that we are medicalising society? That we are really trying to exert control, as in the case of some kind of political ideology, over patients who suffer a little bit? The point is - and Professor Rose also raised it - how much human suffering can we tolerate before prescribing a drug? Look at history, look at the amount of normal suffering - we all suffer a little bit sometimes, but we don't necessarily take drugs for it. But there is a kind of ideological movement which, whether we like or not, is indeed supported by the pharmaceutical industry. Because if you can treat people who are shy with a drug, you will simply sell more. And still I believe that we cannot ignore the fact that, to some extent, we are indeed medicalising human suffering. Whereas not all human suffering, I guess, should be treated in a medical fashion.

Of course, you will treat terrible pain, anxiety, psychosis and severe depression. There is no dispute about this. I myself am not able to define where human suffering begins. But there is a tendency to exert control when someone falls outside certain ill-defined limits. As soon as something is wrong, then you have the right to be treated with a drug. I think this is fundamentally wrong.

PETER MOLNAR: If you know the Diagnostic and Statistical Manual in psychiatry, you know that in order to coin a disease you just have to reach the threshold of significant impairment or distress or whatever. If you can define what causes this distress or impairment, that's a diagnosis. Indeed, I think suffering is a part of life, just like joy. People create a lot of demand for remedies that free them from suffering. Although I think it is impossible to escape from suffering because it belongs to life.

I would like to point out that drugs are a cheap solution. In traditional societies, you find a lot of social capital. They don't need a psychotherapist to discuss issues about love and such things with an unknown person, because they have friends, they have a priest, they have their parents at hand who can help them with such issues. Modern societies are very depersonalised. They don't have that social capital, but they have the money. So the people have to get help from unknown third parties, which they have to pay. And drugs are definitely cheaper than interpersonal contact.

MARJAN SLOB: Do you mean that we are looking for an easy way out with this medicalisation?

PAOLO VALERIO: I want to underline that we can't see what's happening in the brain. With most emotional problems, we don't know why they are happening. So we are working not on etiology but on symptoms. This is a limitation, because technology always offers you new challenges: e.g. I work on transsexualism and, of course, I'm searching for what is best for these people. They want to change their sex, but are they just wanting to change their sex or are they seeking a sort of castration that offers them the possibility of receiving juridical recognition of their new status?

I agree that one of the most difficult issues is how to cope with the psychic pain that comes under this phenomenology. Is transsexualism a new pathology or are those people the discoverers of a new world towards which we are moving? Twenty years ago, homosexuality was considered a pathology. We have to consider the limits of the Western world. For homosexuals, life is different in Los Angeles and New York or in Sicily, Naples or Dallas.

WRYE SENTENTIA: I want to comment on pathologising and on the medical model as authority. I think society would benefit from greater education regarding the complexity of these issues rather than have a top-down injunction as to what is and what isn't acceptable. Homosexuality is a perfect example of that. Or we can look at plastic surgery and how it has evolved - there are people who have extreme elective plastic surgery as an identity choice. I think that, again, this has to do with choice, individual and collective.

PETER MOLNAR: My comment goes back to the question: how do we deal with problems in our society? I think it is perhaps cheaper, and also more attractive, to design a pill than to handle the problems which are rooted in our social and cultural environment.

This creates a link to the responsibility of our society for the environment we are creating for our children to develop a healthy brain. Do we decide that a pill is an easier way to handle a problem because we don't want to go into very fundamental problems of our society? Or do we solve these fundamental problems in another way?

IAN RAGAN: I suppose that, coming from the industry, I should be pleased that people regard drug treatment as cheap, because we are normally accused of profiteering at the expense of the sick. On the other hand, I am not very pleased because I find this idea curious that, because it is cheap and effective, there must be somehow something wrong with it. This reminds me very much of my father, who came from Yorkshire, and who believed that central heating sapped the moral fibre of the nation. I am hearing much the same kind of thing here. So let me come back to the point I made earlier: drug treatments are not considered a cure, they're not the only way. We perceive them to be part of a treatment regimen. They can help. That's all. There's nothing wrong with them for that reason.

KARIN RONDIA: Maybe there is nothing wrong with the drugs but with the communication surrounding the drugs. That communication comes mainly from the industry itself, doesn't it?

PAOLO LUCIO MORSELLI: Two comments. One: I would like to remind you that the limits of normalcy depend on knowledge and culture. It is a relative concept that changes with time. If we go back to the Middle Ages, depressed people were burned because they were guilty of sloth, which was one of the seven sins.

The other thing: to get away from theory and back to practical things, I would like to remind you that most brain diseases are chronic diseases which begin in adolescence or childhood. Evidence indicates that if we start treatment early, the chances of stopping development of the disease are much higher than if we wait until patients are 25 or 30.

KARIN RONDIA: This is a different point. If we talk about chronic diseases, it is something else than drifting away form normalcy and taking drugs for conditions that are not specifically pathological.

PAOLO LUCIO MORSELLI: We should define when certain symptoms are the expression of a starting disease. In many cases, what we hear is "this is an adolescent crisis", which is a big lie. Many people hide themselves behind this and 10 years later the person has a very strong depressive disease which ends up in a manic or schizophrenic episode. This should be kept in mind. I don't say that everybody should be treated, but we have to better define the criteria for a correct diagnosis.

WILLY WEYNS: We are now moving to extreme cases where it is clear that there has to be treatment. For methodological reasons, it would be good to state that there are, on the one hand, very clear-cut situations where there is no discussion, where everybody accepts that treatment is beneficial and, on the other hand, normal people at the limits of normalcy, and enhancement.

KARIN RONDIA: But the problem is that the border is not so clear-cut, it is blurred for many conditions.

WILLY WEYNS: Sure, I know. This is the difficult thing we are debating. And there remains a big issue: where to put the limits and the regulations? But it should not be forgotten that there are very clear examples of good therapeutic treatment.

JES OLESEN: We must also consider who is deciding on whose behalf. People who don't suffer tend to think that suffering is an important and valuable part of life. People who suffer don't think so. They want some treatment. How many of you have not heard somebody saying: "If I ever get disabled and can't move my lower body or lose my vision, I would rather die". But of the people who suffer these disasters, how many do want to die? Few. There is no increased number of suicides among those people. After an adaptation period they live on happily. It is important that normal persons don't impose limits on people who are suffering;

because they are not in that situation, they have a difficult time understanding it.

But with that sort of caveat I have, of course, also realised that we are dealing with a very important problem. We don't want to treat or influence normal behaviour. I just want to mention that the art of disease classification is a scientific discipline. I have worked with headache classification, for instance. You take lots of things into consideration before you construct your diagnostic criteria. The whole idea is to single out something that is not normal from a range of normal things. You can see this in the case of anxiety. We couldn't live if we were not sometimes afraid. We would be killed. But at the other extreme, you have very severe generalised anxiety which is a biological disorder. In my own field, 70 % of you have had some kind of ordinary headache. That's not a disease. But in the same continuum, you have 3 % of the population who have headaches lasting several hours and occuring 15 days a month. The problem is: there is no clear cut-off point. I just want to make you aware of the fact that groups of experts are continuously working on such things: how to set the borders between what is normal and abnormal.

ALISTAIR NEWTON: I am slightly disappointed that the discussion so far has not at all addressed the fact that patients must be involved in helping to decide when the point comes when treatment is appropriate. It has been shown in several surveys that, when scientific ideas are put forward about which treatment is required, very few come within the top ten of the needs that the patients actually express.

MARJAN SLOB: Can you give an example of the needs which patients express?

ALISTAIR NEWTON: For example, if you ask them whether they want a drug which might cure e.g. Parkinson's disease, very few of them are interested in that. They are more interested in bowel function and quality-of-life issues, which are not addressed very much by science. There should actually be a dialogue which brings together both sides of the equation so that the scientists are better informed about what patients actually need and patients are better informed about what is possible.

PAOLO VALERIO: I am a little bit disappointed that psychiatrists and neurologists are speaking about the problems of adolescents - and not child psychiatrists or child psychotherapists. I think that we are lacking the experience of people who really work in this field. I work with adolescents and very rarely do I give them any kind of drug.

We are also speaking about prevention or repression. Are we speaking about how to prevent an emotional problem in an adolescent? Taking care of the family surroundings? Would you give a girl with anorexia nervosa anti-psychotics? Or would you try to understand her family which is trying to cope with the situation, and give help from a psychotherapeutic point of view? I believe that, for girls with anorexia nervosa, drugs are very often useless, but any other kind of treatment could be helpful.

STEVEN ROSE: I didn't want to make it an either/or situation in that kind of sense. What troubles me is the shifting border lines that we see. The world market for antidepressants is now estimated at something like 27 billion dollars a year, and that's increasing. The question that interests me in this context, and that we are not answering, is: what was reflected in a graffito I saw many years ago on an Oxford University College wall? It said: 'Do not adjust your mind. The fault is in reality.'

Is something changing in reality which is actually increasing the psychiatric burden that we see at the moment? Or are we as a society becoming more aware of the problems which have always existed and therefore do we see the need to treat and extend the range of the use of drugs?

Those issues take us outside the terrain of exactly what the citizens' discussion can do about these developments at the moment. But we don't know the answers to those questions. Those are the framing questions within which we have to say: but what is the function, what is the intention of the treatments with the drugs that are available at the moment? It doesn't seem to me that there are hard and fast lines that are always going to be there. If depression is increasing as a diagnosis: why?

KARIN RONDIA: Is there an answer to that?

JOHAN DEN BOER: Not a direct answer, but I wanted to pick up on some of the things that you said earlier. I think that we must not forget that 'what is normal?' changes over time. Things that are acceptable now were sometimes not acceptable in the past. IQ, for example, has changed considerably over the past 50 or 60 years. We are all more intelligent globally now. Secondly, I perceive a kind of tension between the idea that you need to suffer and work to achieve some goal, e.g. in increasing happiness, and the possibility of taking a drug. It is an interesting tension because it is also somewhat paradoxical - to the extent that people say: we need to go to the root of the problem. We need to find out why the person is depressed. They decide that you have to work on it and go for years and years in therapy. That is acceptable because it ensures that we get to the root of the problem. But when you look at what's actually happening, people are very happy to take the drugs. The public is in two minds with respect to this issue, I believe.

MARJAN SLOB: The public is divided.

JOHAN DEN BOER: The public is completely divided, I think. Because there is this ethic that you need to work through the problems, that you have to suffer in order to enjoy a better life on the one hand. But then

when you look at what's happening and the success that the drug companies are enjoying with the wide range of products that they market, people behave differently. Everybody takes Ritalin, everybody takes Prozac, everybody is happy.
The great danger I see there is not so much the use of cognitive enhancers. That's okay, why not? You can send your kids to good schools, you can also give them drugs that will make them function better. I see no problem with that. The great danger is that it might induce a regression to the mean in a way, by preventing people from experiencing a range of conditions, life situations, etc, - not necessarily suffering - but that will make us all more similar to each other. Lack of diversity in psychological states is a danger for society as a whole in the long run. It brings up Aldous Huxley's visions of the future in a worrying way.

WRYE SENTENTIA: Following up on your statement, the question of different approaches to ethical behaviour and relation to the work ethic or religious ethics, I think that these are tensions in society that cannot be resolved. We need to find other models to approach how to reconcile cognitive enhancing as a personal choice, but also at the same time bring the public along so that there is no decision-making in ignorance. The role of the physician should be to provide greater education to the public so that wise choices can be made. In the US in particular, it's looking more and more like the regulation of cognitive enhancement is going to be prohibitory. I think that's a mistake. I would highly recommend you to read the Presidential Council's recent report *Beyond Therapy and the Pursuit of Happiness*. It raises a lot of these issues and comes down on the side of prescriptive happiness: telling people what is the right form of the pursuit of life and what is not. I think that's unfortunate. I would hate to see the smart Europeans go the same way as my unfortunate country.

PETER MOLNAR: A very short comment. I want to suggest a kind of natural experiment. Some of the countries that used to be communist (including my own country) are associated by so-called westernised society with all those bad things. Civil society is underdeveloped in these countries and trust in them is quite low. Knowledge of the potential of civil societies in general is not very developed either. Therefore the danger of medicalisation or even of scientification is there: to see social problems as medical or scientific. The experiment is: if by dramatic changes we manage to improve the state of civil society, will this trend towards medicalisation decrease? Hopefully yes, and then we can conclude that a medicalisation or scientification process is part of development. They are sequences in a process.

You all know that the early interactions of a child are very important. There are studies showing that America is very bad in that respect because of their numerous alternative means of satisfaction. Face-to-face interaction is restricted whereas alternative means of satisfaction are developed: drug use, alcohol (ab)use. American studies suggest that alcoholics, drug users and workaholics are more likely to have suffered relative empathy neglect during the early period of life. Therefore, again, the process view: we can't say anything that is valid for everybody in all generations. But we might say: OK, there is a potential correlation, so let's try to improve the conditions for the next steps of the process.

MARJAN SLOB: So, here again family problems become medical problems and they shouldn't.

BERNARD REBER: I have three points. The first is a logical question: if we have both normalcy and enhancement in the same class, how is it possible to speak about enhancement if we don't know what normalcy is? I'm not talking about deep and profound disease.

Secondly, I propose to distinguish between normal behaviour and abnormal ways of healing. If all is normal in healing, where is the issue?

To finish, a question to Mr Newton: could you tell us if our vote choosing these three questions was in line with patients' concerns?

ALISTAIR NEWTON: In my view they are less relevant than several other issues, much less. I think that these are very heavily oriented towards a scientific view without taking into account the views of the patient.

KARIN RONDIA: Couldn't it be that they are oriented toward the citizen's view, because not all citizens are patients?

ALISTAIR NEWTON: Forgive me. All citizens are patients now or in the future. Even scientists.

KARIN RONDIA: Maybe this brings us back to the beginning of our discussion. Of course, I hear very well what you say about the sufferings of patients. The problem is: when do you become a patient or when are you given the opportunity to become a patient within this growing medicalisation context? Is it normal that everybody is considered as a patient?

ALISTAIR NEWTON: It may not be normal but everybody should be considered as a patient at the moment. You asked when suffering begins. Perhaps this is the wrong way round. Perhaps suffering is not where the needs begin. The needs begin in perception, not at some point that is decided by a scientific view. They begin at the point at which the patient perceives a need. And that need may have no relevance to what the scientist has decided. That's the point. If the scientists are here to provide help for people, then they are trying to provide help at this point of need. If you are not going to take into account what the patient perceives as the point of need, than you are not serving those needs.

KARIN RONDIA: What determines the needs of people? We were talking about shyness - when do you become a patient because of excessive shyness? Who decides? You? The physician? Society?

JOHAN DEN BOER: Let me give one simple, practical example. Maybe this will help the discussion a little bit. My wife is a psychologist and she does psychotherapy a lot. And she sometimes treats married couples who don't get along very well and who intend to separate. They are all, without exception, very unhappy. What do you think they get from the general practitioner? Antidepressants. Why? Because the woman cries. Is it good or bad to prescribe antidepressants in this situation? I mean, these are no severe mental disorders, we are really here within the limits of normalcy. My statement is: this is wrong. The general practitioner should not prescribe antidepressants for marital problems. The reason why the general practitioner does this is that the message of scientists and of the pharmaceutical industry is that you can elevate your mood in all kind of circumstances by simply prescribing an antidepressant! What this general practitioner does at this point is to medicalise relational problems. Here we have a clear-cut demarcation criterion to state: this is wrong behaviour. In my view it is wrong.

KARIN RONDIA: What if the people themselves ask for it? If the woman says: I am unhappy with my husband?

JOHAN DEN BOER: Look for another one!

KARIN RONDIA: Yes, but that is difficult. It is much easier if you prescribe me some Prozac!

JOHAN DEN BOER: Yes, but I think it is looking for solutions in a direction that is completely wrong.

KARIN RONDIA: But isn't it what happens?

MALE VOICE: I want to point out that not only is depression increasing, but also the diagnosis of autistic disorder. People speak about autistic epidemics. I want to point out another interesting finding. As you all know, twin concordant studies were undertaken to determine how much genetic or environmental, non-genetic factors, contribute to disease. In the forties, there were several studies on a genetic disposition to smoking. In males, the genetic predisposition was around 40 % and in females 0 %. Now, in the last studies, the females have exactly the same genetic predisposition! This can be taken as proof that social norms act like a buffer for genetic predisposition. Are we treating diseases that are caused by changing society, changing cultural values? Maybe we should not jump to the drugs but think about how culture is contributing to a disease. Maybe there is a refrigerator society that is behind e.g. increasing autism?

ALISTAIR NEWTON: I will have to leave this meeting now, and there is one last thing I would like to say. I think we should all say 'thank you' to the industry for all they do. 93 % of the drugs which have been developed were developed by the industry. That's a huge input into what happens.

This morning I mentioned economic imperatives, but there are also societal and political imperatives which we all have to live by. This is best expressed by what the European Research Commissioner, Philip Busquin, said at one of the meetings: 'The dialogue between science and society and industry is the only way forward'. We all have to work together; politicians expect it.

There are clinicians and basic scientists in this room. The clinicians see patients and have a better opportunity to understand what the patients' needs actually are. Very often patients do not understand what they actually need, so sometimes the clinician will have a larger weight in the balance between patients and doctors. But basic scientists should see that they need to consider carefully for whom and for what they are working. I ask you to consider that today.

Today you will talk about societal issues. 'Limits of normalcy' to me is not a societal but a basic science issue, and you spent an hour talking about that. I hope that for the rest of your discussion you will bring more societal issues into the equation.

Plenary discussion of 'Reductionism'

WITH AN OPENING STATEMENT BY PROFESSOR DETLEF LINKE

MODERATOR MARJAN SLOB: The topic of this plenary discussion is 'Reductionism', or 'mind equals brain'. So basically we are going to solve the classical, philosophical mind-body problem in the next hour! The catch phrases we assembled under this category are:
'mind = brain = molecules', 'brain imaging and diagnostics', 'ideology', 'indeterminism', 'predicting the future', 'human = social animal', 'brain research becomes cultural narrative/ the brain is the new cosmology/brain becomes metaphor for what it is to be human, hyperknowledge', 'no more art', 'no more mystery' and 'brain in society or brain of society'.
We invited Professor Detlef Linke, philosopher and professor of clinical neurophysiology, to start this discussion by presenting his views on the topic.

DETLEF LINKE: If you are looking for rodents placed somewhere in the world, you can detect where they are by looking at the hippocampal neurons active in their brains. We all have a system of navigation in our brain and we always place ourselves somewhere. This is a basic ground for cognition: to tell ourselves where we are.

We are all looking for a place where we are safe - and that place is the brain, well enclosed in the skull. This is our 'insula' on which we are safe in our permanently changing world. But we should remember that the brain is not an axiomatic system. We cannot do any mathematics with it. I am on the level of hippocampal attribution, of just giving a place to cognitive things - a bit like the way you are putting our quotes on papers on the wall today. Brain research is not about abstracting and finding last categories, but just a way of summarizing facts and findings. That's the current level.

We should note here that it is not possible to have an axiomatic system in this world without contradictions. Even mathematics is no longer the ideal it once was. We have come to undecidabilities in systems and these undecidabilities are probably infinite.

The infinite comes into our mind when we say: 'outside is infinite information, within here is a constructive system which reduces, so everything can be safe'. This process of reduction is made by the brain. But if I naturalize this cognitive process in the brain, I will end up with the same infinite nature as outside. This gives some movement to the whole story.

So epistemologically we are in a situation where we have to ask which complexities to choose - since we cannot start to investigate what we think to be secure. The question is: is just producing new research based on our current model of the brain the solution to finding a way out of the labyrinth of information? We are in the same situation as information scientists developing new networks. We do not have to cope with our problems by always using the same network. We can also change our network.

The question of reductionism is no longer the situation of finding the right axiomatic system from which to start. It is rather a way of properly handling complexities. Usually reductionism is attributed to natural scientists only. If we take the question of reductionism, one might as well say: 'Isn't it reductionism if you want to reduce all theses complexities to some ideas of mind or consciousness?' So, the great question is whether there are no simplifications on the other side too. If you are doing the equation mind=matter, that is, if you are putting forward a 'monism' of mind and body, than of course the question is: 'well, if it is monism, do we perceive it as mind? As matter? Or is it both, and if so, how is it mixed?' The question is open, it is not decided. In the equation 'mind=matter', or 'we are the brain', it is an open question what is affected more by this statement: we, or the brain.

The great questions begin with the tradition which poses a dichotomy of mind and matter. We should overcome this dichotomy, because the brain is not just matter, certainly not. If you stick to the old traditional dichotomy, you will have to be aware of the fact that you will not find anything else in the natural world which has so much form as the brain. In a way, the brain is much more form than matter. And it produces these forms out of itself. So the matter-mind dichotomy is quite problematic and we should look to overcome it.

But not by a simple equation, by stating 'mind=brain', because there are some things hinting at beyond the brain. Let me refer to two topics. Neurosemantics is one dimension. The meaning of words does not only depend on the brain processing it, but also on the biography, on the history of the processing body and the history of the world it finds itself in. The meaning of H_2O depends also on the history of ideas about H_2O. So it may be perceived as a chemical substance or as an element with a history associated with fire, earth and air. These are quite different ways of looking at things. Processing ideas and meanings can open up our brain to very long histories of evolution and cosmology.

The second example aimed at showing that the equation 'mind=brain' is too simple as it does not deal with the fact that the software (mind) can change the hardware (brain) itself. Should we always stress that there is a hardware with which our software is working? I think everybody working with computers knows that it is not a very effective software which always describes its own hardware. In some sense, the dichotomy is there again in the way in which we differentiate and say: 'I am not my brain'. But it is not sufficient to say, as some philosophers do: 'Because I can abstract from myself, get into a distance from myself, I cannot be my brain'. The interest of the brain scientist is in how self-distance can be realized by a brain process too. I think there are great opportunities for brain theory.

AXEL CLEEREMANS: I would like to connect with that in stating that we have to be very careful how we communicate these ideas to the public. Two things. One: if we are to have a scientific approach to the mind-brain relationship, then we have to assume that you can do reductionism. Yes, the hardware is the software. The mind definitely is the brain, in the specific sense that there can be no mental event that is not also a neural event. We have to be very clear about that idea. The mind is the brain.

But this is only half of the story. One of the most interesting developments in recent brain research is precisely the fact that it is beginning to show now that the very structure of the brain is influenced by cognitive developments, social interactions, language, culture etc. We find traces of the brain's constant interaction with its environment and with the culture in which it is embedded in the very structures that support our mental life.

This is what is exciting. Yes, there is total reductionism on the one hand, there is nothing in the mind that cannot be reduced to brain activity. This I believe is the basis for any scientific approach to this story. But it is just not any old reductionism, it is an interesting sort of reductionism in which you have different levels of description. The relationships between the levels of description are of interest. There might emerge something at some level-plus-one that you cannot reduce to a simple one-to-one correspondence with what is happening at the lower level.

These are the two key ideas we have to communicate to the public. In part to alleviate fears in the mind of the public that if we say something like 'you are just your neurons', free will and personal identity will evaporate, degrading a human being to just a bunch of neurons floating around, that is to zombies. We have to make clear that in a sense this is true, and in another sense it isn't true. And this is difficult to communicate.

KARIN RONDIA: This is an important topic, and we are talking about communicating with the public. So, can you please give an example of how the mind keeps track of the environment.

AXEL CLEEREMANS: The discovery of the mirror system is an important development. The idea that there are neurons that respond specifically to an action that you perform yourself as well as to an action that you observe but that is performed by somebody else. That is a crucial link between the very manner in which the brain is organized and the fact that it is embedded in a social environment. Many other examples show this close link between the social system in which the brain is embedded and its very structure which is continuously shaped by experience.

MARJAN SLOB: I see some similarities between the comments of Detlef Linke and Axel Cleeremans: you both say there is no ontological division between mind and brain, as we like to phrase it as philosophers. The mind is the brain in an ontological sense, in the sense of what is. But that leaves open the question of how we want to assess what is.

Professor Linke said we have to choose our epistemic systems. That will be a crucial question, because - as Berit Faber said - the public is afraid that the soul will be gone, since they now locate the soul in the brain. And if we define the mind as the brain, they don't see a place for the soul. This is what we have to address.

JOHAN DEN BOER: I agree with the points that have been raised. The brain is not only embedded, but also 'situated'. This situatedness is important in the current philosophy of the mind, and even in robotics. But still I think as a neuroscientist.

MARJAN SLOB: Can you explain the difference between 'embedded' and 'situated'?

JOHAN DEN BOER: 'Embedded' means that the brain is a part of the body. It cannot be dissociated from the body since there are a tremendous amount of biological influences in the periphery which have a direct influence on the functioning of the brain.

Secondly, you asked for an example of 'situatedness'. It's very simple: when you ignore a child at a very young age, there is a very strong impairment of brain development. You can see it not only in terms of microbiological changes, but also by doing an MRI: you will see a smaller brain. A brain simply does not develop when it is not situated.

As neuroscientists we do ask reductionistic questions. I know only one holistic question that was asked by Sigmund Freud: 'Was will das Weib?' (What does the woman want?). That's a holistic question which hasn't been answered yet, not even by neuroimaging, I guess [laughter]. An important point is that a lot of modern neuroscientific research conveys the wrong message to the lay public. If you look, for instance, at the work done in terms of decision-making, then you get the impression that parts of the brain are involved in reward, in emotional processing, in cognition of course, and in decision-making. This is what philosophers call the 'neoneurological fallacy'. The fallacy is that you are attributing psychological capabilities which only human beings possess to isolated parts of the brain. This is a complete mistake. It is incoherent and confused. We as neuroscientists should warn against this excessively simple picture of what the brain is doing.

MARJAN SLOB: Why is it incoherent and confused?

JOHAN DEN BOER: It is confused because a small part of my brain does not take moral decisions. I am the one who is taking the moral decisions, not a part of my medial prefrontal cortex. It may mediate the proper functioning which in the end leads to me, as a whole person, taking the decision. But that is completely different from saying that the medial prefrontal cortex takes the decisions. That's wrong, that's a fallacy.

Sometimes neuroscientists are to blame for that: it's nice to have your name in the paper and so you present a breakthrough about a part in the brain that is taking the decisions for you. But I think that is completely wrong.

KARIN RONDIA: So is it there that we put the soul?

JOHAN DEN BOER: I think that this is asking the wrong question. It is wrong to ask where the soul is seated. I think it is also a wrong question to ask where consciousness is.

KARIN RONDIA: It's a wrong question, but I mean it in a provocative way. If this is what the public wants to hear, we have to explain to them why it is the wrong question and what is the answer to their questioning, even if the question is wrong.

JOHAN DEN BOER: But that's a philosophical issue. Now, there is a trend to naturalise consciousness, and rightly so. But the question still is: can the brain sciences in the end fully explain what consciousness is? I tend to say 'yes', but I still hesitate a little bit. Because by just looking into the brain, you will not find the answer; you are ignoring the fact that the brain is situated. So you're looking in the wrong place.

AXEL CLEEREMANS: I think there is no soul.

JOHAN DEN BOER: I agree.

AXEL CLEEREMANS: So I think it is wrong to say that 'you' is located in a particular part of you. But you're still your brain, in the sense that there can be no mental event that is not also a neural event. It doesn't mean that you can do a simple one-to-one relationship between particular mental functions and particular brain regions. I agree with you that's wrong. The challenge is to understand how the hardware can be shaped by the software, that is: the social and cultural interaction, the development, etc. The challenge is to understand how the situatedness of the brain can be embedded in its hardware.

MARJAN SLOB: Let's ask Berit Faber what she means by the soul.

BERIT FABER: I didn't express my own idea. I tried to distil what my feeling is about the general perception among the public, and where maybe some of the fear stems from. I don't think I'm qualified as regards characterising what the soul is!

LEO ZONNEVELD: Brain science looks at and rethinks the future of the human phenomenon. It expresses a wish for greater change, for greater self-expression, for greater self-improvement. You can see it as an evolutionary trigger.

Developments in the brain sciences are not only about medical innovations, we also see a new cultural, virtual domain being created. This domain will be populated by biological organisms, but also by silicon microchips: central nervous systems (CNS) that are being built in Britain at the moment, using revolutionary algorithms. What we need is a strong feeling of identity within a number of virtual worlds that are now being created.

What might be very important to do is to see whether the present developments of new scientific concepts help us in improving the revolutionary authenticity and identity of the phenomenon of man. I think brain science can be a help in stimulating us in that particular direction. As long as we are being encapsulated by seeking the soul or seeking the brain or the mind in itself as a concept, we won't be getting there. We need to open up to what the human phenomenon really is. What we want to look at and seek for is a new way to see, express and help the human identity.

ANDREAS ROEPSTORFF: I think you can adopt another perspective: What if the mind equals the brain? You can adopt the perspective that all the phenomena that we know exist in human minds somehow have to exist in human brains. In other words, rather than reducing the mind to the brain, the idea is that what we know is taking place in humans has to take place in brains.

It's not a question of asking questions from within brain science that give us new knowledge about the mind. The trend seems to be a different one: that there are a lot of ideas about what it is like to be human, what communication is doing, what emotions are doing, etc. It turns out that one can ask interesting questions in brain science that throw a light on these issues again. This is really where the interest in developments in cognitive neuroscience comes to the fore now.

Note: this is not an issue of reducing things to the brain but rather a matter of saying: 'We know a lot about what goes on in the mind, and we are all experts in that.' If you want to accept the equation mind=brain, than you have to say that 'what is human has to be found in the brain.' That allows us to ask some questions.

My starting slide demonstrated a woman looking at a computer screen. She was a brain researcher looking at a brain image. The image she was looking at was a representation of her own brain. I think you couldn't say that she is that brain. In constructing the brain sciences, we have created a whole new inscription. We have added a new layer of descriptions to a terrain in which there is now a tension between our knowledge of ourselves and of one other. We will start seeing tension in relation to some kind of new knowledge coming out of what we know about the brain. This is not a matter of reductionism. This is a matter of new inscriptions and new types of knowledge. I think we know more about the mind than we do about the brain.

DETLEF LINKE: I want to pick up on this question of the soul. When talking to the public, we are not reducing the soul to the brain. In my opinion, nowadays it is much more the concept of the self which for the public has replaced the soul. Brain science for the public rather refers to the self, and thereby brain science becomes understandable for them, because they say: 'These processes, that's me too.' That's very fascinating, but what will it do to our concepts of freedom and the like? It's not so much the problem of determinism which shocks many people. The problem is deeper. It is the fact that the concept of freedom might be endangered by saying: 'you are your brain'. Because freedom was formerly defined as being against natural processes. Freedom is no longer in the forefront of interest now. That may be the real problem.

FLAVIO KELLER: I do not think the hardware-software metaphor is a good way to approach the mind-body problem. I want to remind you that Freud used concepts like sublimation which comes form chemistry, which was the most technological and developed part of science at that time. Any metaphor is conditioned by the technological state of the art of the moment. So now we use the computer. We should get away from this simplification.

We should remind ourselves that our purpose is to get out to the public. Many people still use the concept of 'soul'. As scientists we cannot get out and say: 'There is no soul'. The soul has philosophical, religious and other implications for many people and you cannot just say that these lack any basis. It's not respectful. We are not entitled to this kind of affirmation on the basis of our knowledge.

Every discipline works with certain methods. We are using the methods of chemistry, physics, electricity, whatever, and we can capture a particular part of reality. But there are also other methods for approaching reality that are not related to these sciences. All the time we speak about personal self, history, and the like, we are in the domain of other sciences that don't use these methods. Their content is also real.

PAOLO VALERIO: Can we say that the brain has to be embedded and situated? Situated in space, time and relationships. The brain not only needs to be fed, but also to be loved and taken care of. A brain that is well fed but not loved breaks down, as was shown by research on babies left alone after the war. The brain is not a computer, since it does not only feed on electricity. You need to take care of it, give it special care.

Also, not every brain reacts in the same way. An event may be a big trauma for one person, whereas for another person it isn't so at all. Everybody works through their experiences in their own way. This is also a very special quality of the brain. It makes it different from the stomach, the liver and everything else.

In fact people's experience of the world is highly subjective. Even elementary perceptions like, for example, sight and sounds are not passive processes because we process actively any incoming stimulation, selectively focusing on some aspects of the stimulation and ignoring others. Once we have focused on these aspects, we organise them and make our perception personalised and subjective. So people often see what they want to see or what they expect to see. And it is clear that motives and expectations often colour human experiences.

ZOLTAN NEMES: Yesterday the question came up of why people don't give us brain scientists the money we need. We seem to have an answer now: they know that we deny the soul. People feel that they have a soul, freedom, love and whatever other nebulous categories which we cannot tackle with our cognitive instrumentation. They would rather spend a lot of money on soul, love, freedom, and so on. We cannot approach it. We are just working for ourselves, to amuse each other, and not for the lay people who have soul, love and freedom!

Plenary discussion of 'Responsibility and free will'

WITH AN OPENING STATEMENT BY PROFESSOR JOHAN DEN BOER

MODERATOR MARJAN SLOB: I see a fluent transition from the topic of reductionism to the problems relating to freedom and responsibility here. Several of those present have mentioned that the public is afraid because free will and responsibility may get lost if we treat ourselves only as matter. Let's use this as a starting point for the discussion in the third workshop on responsibility and free will. Professor den Boer wants to make the opening statement on the discussion of 'Responsibility and free will'. Johan den Boer is Professor of biological psychiatry at the Department of Psychiatry of the State University of Groningen in the Netherlands.

JOHAN DEN BOER: Not so long ago, a lawyer in the US defended a serial killer by simply stating that he wasn't responsible for his acts for the simple reason that his 5 hydroxy-indolic acid concentration in the cerebral spinal fluid was too low. He referred to a tremendous amount of biological research showing a significant correlation between a low concentration of this metabolite and impulsive behaviour and outward and inward direct hostility. So he pleaded 'not responsible' because his client showed a biological abnormality. If this is the message that is being conveyed from neuroscientists to the lay public - lawyers are lay public - then I think we are conveying the wrong message. This is simply wrong and it is a very relevant issue because there has been an increase in violence. In psychopathy, for instance, there is something really wrong in the brain. And I thank God that there is something wrong in the brain of psychopaths, because otherwise evil people would really exist. Imagine that as normal! Still, the fact that there is something wrong in the brain of serial killers does not lead automatically to the conclusion that they are not responsible for their deeds. That would be the wrong conclusion, in my opinion. I would opt for at least partial responsibility. The issue of responsibility is tricky, because it is related to questions of justice and law. Denying responsibility on the basis of a functional impairment would not be correct in my opinion.

JAN STAMAN: You have to take responsibility for the self that you are. For being coherent. That is a life-long job. It is a dialogue you have with yourself. If I hit someone on the face, because my metabolic substances of one type or another are too low, I would not feel happy about that afterwards and I would try to figure out why I am reacting in this way. If I am brought to court and the judge asks me: 'Why did you hit someone again?', and I have to say to the judge: 'I try to stop doing it, but there is something in me which is beyond my control', the judge will immediately ask: 'Did you seek help?' And if I did, if I am attending three psychiatrists and taking pills, but it still does not help, yes, then.

KARIN RONDIA: .. You might try neurosurgery!

BERIT FABER: Perhaps I am interrupting something, but when talking about responsibility I think we should also introduce the notion of transparency. If we strive towards transparent procedures for scientists and the industry as regards the way we decide about things, then this might help in facilitating responsibility and giving the public the right information.

JAN WILLEM BERKELBACH VAN DER SPRENKEL: We scientists try by nature to reduce complex questions to simpler ones in order to answer them. Sometimes this leads to oversimplification of the problem. Sometimes we want to present our results as if they have much more impact than in fact they have. We use statistics to enable us to say: 'this is indeed something that might help us and our patients'. Statistics give body to the answer to our scientific question. But it's not only statistics that should concern us; it's also the impact of what we are doing. We are not only giving a statistically significant answer to the public, we also have to acknowledge the impact of the answer that we give.

Especially - and that's where Mr Alistair Newton comes back to my mind now - we as doctors know we should talk to patients. We listen to their problems, try to answer their questions, try to constantly listen to the problem that this individual patient has raised in his mind. The contact between doctor and patient is very close and the answers are very much tailored to the question.

In science in general, I think there is a very big distance between practitioners and the public, so big that we do not reach the public any more. We should convey to the public that we are using simplifications to answer our questions and that it is not the complete answer. I think that will help to improve the image of neuroscience. It also helps when you are a neurosurgeon to tell your patient that you can't solve all their problems and that you don't know all the answers. People are always very grateful to hear that.

PAOLO VALERIO: Speaking about moral values: as psychiatrists we run the risk of becoming the watchdog of society. If parents bring a teenager to my office who is very promiscuous sexually speaking, what should I do? Just say: 'Don't do this'? Or should I try to pay attention to what's happening? To engage in a sort of active listening means trying to give explanations, not justifications. What's happening to a paedophile? He's doing something very bad from a moral point of view, but from the psychological point of view there can be an explanation for his behaviour. And this means also helping the public to understand that explaining and guiding do not mean justifying. This is one of our tasks.

A neurologist can give you an explanation for the temporal fit of an epileptic. For behavioural problems we very often cannot easily give an explanation. We can perhaps give an explanation, but only through a long diagnostic process that entails getting into a relationship with the patient and via this relationship trying to understand something more. If we consider that person as an isolated patient with a brain problem, we really run a big risk of becoming watchdogs. We should also consider the network of this patient's relationships and his history of emotional problems.

DEMOSTHENES AGRAFIOTIS: Who is the owner of the brain? We, nature, God? If you put an object - let's say, a brain - in the social arena, different groups, professions, discourses come there trying to claim and capture this object, to define its limits, to give it dynamics. It's some kind of conflict. Who is going to define the area of brain? Who is going to give the best definition, who is going to analyse, manipulate, make the rules? This is a very important question in knowledge societies.

MARJAN SLOB: It's the epistemic conflicts again, I think.

DEMOSTHENES AGRAFIOTIS: It's not only epistemic. An object, let's say the brain, is open and closed at the same time. Sometimes we have some kind of stabilisation of discourses, of practices around this object. There is also some kind of destabilisation. The question is: who brings this destabilisation about? Sometimes it stems from knowledge, sometimes from passions, sometimes from Europe, from events, the arts. Sometimes a crucial factor destroys the stability.

To me the question is: is brain research today capable of destabilising discourse about the brain or are there some other factors outside brain research that configure the whole discussion? The industry maybe? Who is the strategic actor?

MARJAN SLOB: So you are pointing to the interests behind the definitions?

DEMOSTHENES AGRAFIOTIS: I am saying: if you don't place the question of responsibility within this field of dynamics, it will become very abstract. Whereas if these dynamics are clear and well understood, you can start to think about how to organise these dynamics in order to prevent one group from keeping this territory as its own.

WRYE SENTENTIA: Ownership of the brain is an important term. Serial killers were mentioned. Maybe there is a habit to crime; neuroscience might enforce that idea. I wanted to throw out the question: to what extent are we our brother's keeper? Because in some sense the issue of personal autonomy and choice must be found to intersect with the revelations of science. Yet at the same time, the legal parameters of freedom may come into conflict with them.

I just want to mention a particular case that was very disturbing in the US recently. It is the case of Charles Singleton, a man on death row, who was set to be executed. However, he was mentally incompetent. He was forcibly given an injection to make him sane so that he could subsequently be injected with a drug to kill him. This ludicrous social responsibility in relation to personal autonomy and choices is really vexed and I'd like us to explore that more.

FLAVIO KELLER: I wanted to come back to an example given by Professor den Boer, the one of the killer defended by his lawyer who said that his client killed a person because his CSF 5 hydroxy-indolic acid was too low. I think the key word here is 'because'. It brings us to the more philosophical issue of causality. If I say: 'The man killed because acetylcholine was released from his neuromuscular junction', that's OK. It's true, since that brought the muscle to contract. If I say he killed him because his CSF 5 HT level was too low, it's also true, but it's a causality of a different type.

In the first case, there is a deterministic relationship because it is known that the impulse transmission form nerve to muscle is a one-to-one. The second case is not so clear. If we go further and say this man killed the other man because he wanted to steal, that's also something that is true.

What I want to show is: we use this little word 'because' for very different things. The issue of causality is a big one. And it has enormous societal consequences. Because freedom implies responsibility and vice versa, and I am responsible when I can claim that I am the cause of my own act.

ANDREAS ROEPSTORFF: I think, in particular, that this issue of responsibility and free will is probably one that really lends itself well to public discussion. Much more so than the discussions of reductionism and the brain, because they are technical. Berit Faber made this beautiful comment yesterday. I think she said: 'If you actually ask lay people which kind of things you are allowed to play around with, I think that they would say: "It's OK that you repair a heart, that you repair a muscle. But once we enter into something like cognitive enhancement, people might think that this is a no-go area."'

Although you are well aware that some kind of pathological behaviour might be coupled to deficiencies in the brain, there seems to be an autonomous level of discussion that has to do with personality and moral issues and simply shouldn't be reduced to something else. I think this level of discussion is a level on which you can very easily discuss with the public, as some of the work done by the Danish Ethics Committee proves. You might comment that neuroscience does not have anything to say about questions of personality and the like, but to me it is quite clear that these discussions of the public regarding moral issues have a lot to say about which kind of implications should be taken from neuro-

science, which kind of knowledge should be used here.
So if you're looking at a target area for public discussions I would say that among those three on the board 'responsibility and free' will is probably the most important one.

AXEL CLEEREMANS: I'd like to remind you of the work of Daniel Dennett, the American philosopher, who in 'The Intentional Stance' made exactly that same point. Dennett is a reductionist who says: 'yes, the mind is the brain, and so in principle we can predict what you will do next, based on my knowledge of the state of your brain and of the environment at this particular point in time'. In his perspective, we are completely deterministic persons and in a way there is no free will.

Except that it is a lot more complicated. It is a monstrous issue really. In a sense there is no free will. Yet his conclusion was exactly the one you are proposing: though we might have this knowledge and we might believe there is no free will, as scientists do, we should nevertheless behave as if it weren't true. Otherwise there is no sense in which the society we live in can continue to function. We should continue to ascribe causality, personal responsibility etc. to a level of description at which there is free will.

ANDREAS ROEPSTORFF: You can also relate such discussions to the point made earlier by Steven Rose: if there is a discrepancy between the mind and the natural world, maybe it is the world that needs updating. I think Dennett is a nice case. There is a discrepancy between common intuition and theories, yes. But maybe it is the theory that needs some updating.

Closing session

Plenary evaluation of the outcomes of the first European workshop 'Connecting brains and society'

MODERATED BY JAN STAMAN

JAN STAMAN: Do you think that this workshop provided us with what we need for working with the public? Can we start up the public debates in the ECD programme?

GERT VAN DIJK: It's going to be extremely difficult to have a public debate about these topics. Some of the issues seem way too specialised for a general audience. It seems that the real work only comes now: to put it into a perspective that is accessible to the public.

GRAHAM FARMELO: I am very optimistic. I think some very pertinent points have arisen that we, in a journalistic capacity, can work on with the public. We must form a bridge between the experts and 99 % of the people. It has emerged that reductionism as a hook to the public is no good. But as we talk with them, it will become a very prominent issue.

JAN STAMAN: Did multidisciplinarity work?

GRAHAM FARMELO: Yes, I think so. It was a successful experiment.

JAN STAMAN: Can lay people get into a discussion about these issues?

FEMALE VOICE: Yes, difficult but do-able. Maybe lay people want to discuss completely different things than experts. We should leave room for people to express their thoughts and what they feel to be relevant.

PAOLO VALERIO: During this meeting, I have learned by direct experience how difficult it is to work in a multidisciplinary staff group. My first language is Italian, and my 'disability' in English is a handicap if English people do not help me to cope with my 'disability'. In order to have a real encounter, I have to improve my English, but the others have to consider my level in English and to speak in an understandable way to me. This means that we have to find a reciprocally shared level of knowledge.

This is true not only for languages but also for scientific knowledge in psychology, psychiatry and neurology. This is the big challenge because very often psychologists speak their own language, without knowing and taking into consideration the language of psychiatrists and neurologists, and of course others do the same. Therefore in a multidisciplinary staff group the first aim is to find a shared language and a shared level of knowledge.

JAN STAMAN: So it feels good to feel like an imperfect person.

DEMOSTHENES AGRAFIOTIS: This is exhausting work. Participation is tiring. The most important thing is: how to organise this kind of exchange? Do you put emphasis on differences and conflicts, or do you try get to a consensus and move ahead? It was not so clear what happened in this group.

We have done the same work for biotechnology. We took 18 months to arrive at some categorisation of the problems. We held 9 meetings in 9 different countries. You are asking too much by cramming this into one day. If you pass this to the Commission and I have to be an evaluator, you will have problems.

General Conclusions and Close

BY GERRIT RAUWS

I will conclude, but I will not draw conclusions. That would be an anticlimax.

The mind matters. We are even more convinced of our approach and issues than we were 24 hours ago. Beyond all the fascinating scientific and technological discussions we had, one of the main ways in which this project has been argued is: the brain has become a metaphor of what it is to be a human being. That seems to me a fascinating challenge for participatory technology assessment.

Of course, for a public discourse, brain science is different from nuclear energy. But we shouldn't be afraid. The first reaction to any topic whatsoever is always: 'It's much too difficult'. There should not be topics which are too difficult for such a dialogue. And there are not. This is what our consortium can offer: to the scientific world, to the pharmaceutical industry, to patient organisations, to the policy makers (whom we didn't invite), we can offer them expertise on dialogue and communication. Expertise that avoids the classical pitfalls of communication and vulgarisation. Ways of communication and dialogue which go beyond the ideas that it is all the fault of journalists and the media, of teachers in schools, of scientists who are not capable of communicating in a proper way. We try to go beyond disconcertations and create a real dialogue.

You, the invited experts and stakeholders, have offered us a lot during these 24 hours. You have inspired us in mapping the issues, in prioritising, in framing the issues from very specific to very broad topics. The output from these two days is not the immediate input as regards what we are going to do. But it is an enormous help in translating and communicating it to all citizens.

Twelve institutions form a partnership. We want to keep in contact on a regular basis with all of you. We want to make a contribution to public deliberation on brain science and brain research.

We are preparing the submission of a call. We have a lot of resources, we need more and we hope to get this from the European Commission. We ask you for your moral support. If you think this is a good idea, express your support for it. But even without support from the Commission we are committed to going further.

I thank you all for your extremely valuable contributions!

Part 2

| Synthesis Report |

BY KARIN RONDIA AND MARJAN SLOB

Introduction

Undoubtedly, ongoing developments in the brain sciences will partly shape the way European society will evolve in the coming years. How and to what extent they will change our ways of living can, of course, not be foreseen with any certainty. But this lack of certainty does not free us from the task of anticipating the possibilities and dangers that these changes might bring. As a consequence, we might have to shape national or European policies to facilitate developments. Or design laws or behavioural codes to avoid the pitfalls of others.

Citizens should not be left behind in these decision-making processes, since the decisions will affect their daily practices and prospects. That is why, in the Science and Society spirit, the European Citizen's Deliberation (ECD) consortium has decided to focus on the societal and ethical issues relating to the most recent developments in the rapidly advancing field of the brain sciences.

This report aims at tentatively mapping these societal and ethical implications. It is based on the statements made and exchanges carried out by a selected group of scientists, cultural thinkers and stakeholders, all working at the cutting edge of the brain sciences who were brought together during a two-day closed workshop in Amsterdam in April 2004, as the launching event of the ECD project (see Part 1 of the present report). These experts were specifically asked to point out the developments in their field which could potentially, in their view, affect society the most. This part of the report is structured along the lines of these discussions but it is also fed by a thorough review of the international literature.

Demosthenes Agrafiotis: "When we speak of a great revolution, like the Darwinian one, or like the current revolution in brain research, what will be born from it?"

In the first section, *Limits of Normalcy*, we analyse the two frontiers of what can be considered a 'normal' state. The first frontier of normalcy is in relation to illness, and it is nowadays a very shifting line because of earlier diagnoses but also because of a growing trend towards medicalising behaviour and personality traits that were not previously considered to be pathological. The second frontier is in relation to enhancement of natural performance; this frontier is about to be crossed by the numerous endeavours to augment human functions. This opens the way to the second section, *Personal Identity*, in which we question the impact which these enhancements and modifications can have on our own feeling of being ourselves. Our brain is what makes us build our own identity... and what ultimately defines life and death. These concepts are also developed in this section.

The third section is devoted to the fundamental issue of *Reductionism*. Beyond the problem of clashing vocabularies, reductionism is a recurrent reproach and a philosophical questioning of the brain sciences: are we what our brain makes us? Or are we 'more' than that? This questioning is not a mere reflection on metaphysics: at a time when brain function is on the verge of becoming the new metaphor of what it is to be human, the common acceptance of these concepts can influence the way we see ourselves and others in society.

The next section, entitled *Responsibility and Free Will*, naturally ensues from the preceding one: if the brain decides our thoughts and actions, what is left to our free will? And to what extent can we be held responsible for what we do and think? Are we predetermined? The old Nature versus Nurture debate finds a new perspective in the field of brain science.

Social Justice is a section that addresses the distribution of the new benefits of brain technology, but also the potential consequences thereof: will we be running a rat race for better performance? Will we be free to refuse it? Will we reinforce the conflict between generations? And finally, how can one reconcile the interests of individuals with the solidarity of society? This leads us to the vast question of *Privacy and Personal Freedom*, in its different declensions, namely freedom of thought, deviant behaviour, the judiciary and law-and-order aspects, commercial and military interests, the influence of the media, manipulation of the mind and freedom of access to information.

Ethics of Research and *Dialogue, Information and Exchange* are the two last sections, dealing with the 'epistemic and semantic duty' to establish clear and honest communication between scientists and the public, on the experimentation and clinical side, as well as on the information and education side. The necessity for interdisciplinarity embedded from the very beginning in any brain-science research is here underlined as a component of ethics itself. Special attention is also given to animal welfare. Regarding information, the accent is put on the need for a two-way dialogue between the scientific world and the public, but also between the industry and the public, and between governments and the public, since so many more interests are at stake than just information.

This synthetic report does not pretend to be exhaustive. That would be a preposterous claim, considering the explosion of scientific knowledge of the brain now taking place, the tricky conceptual issues which are related to the brain sciences, and the exciting but unfamiliar interdisciplinarity of the field. Still, we hope it will serve as a useful stepping stone in thinking through the societal implications of developments in the brain sciences.

Limits of Normalcy

Medicalisation

Man has always been fascinated by the idea of pushing back the limits of his human condition. From the morning cup of coffee to sportive doping, from Viagra to hallucinatory drugs, all available means have been explored, and many of them have been included in our everyday lifestyle. Nowadays, new developments in the brain sciences offer the possibility of broadening this offer considerably. Of course, the first to benefit from this would be patients suffering from a decline in their corresponding functions. But the step from treating a patient to enhancing a normal function can easily be made.

If various forms of brain enhancement (cognitive, emotional or behavioural - pharmacological or not) become commonplace, the baseline of normalcy will shift, raising numerous issues about individual ethics as well as about societal values and norms.

What does it mean to be normal?

Before we can affirm that the limits of normalcy are shifting, we have to know what constitutes a 'normal' state. Normalcy has both a statistical and a normative meaning. There is a wide variety of ways of being normal, some of which are statistically abnormal. One can, for example, be abnormally tall or have an exceptional musical ear. Atypical people, i.e. abnormal in that way, can happen to be as functional as typical ones (or more so).

The *normative* concept of normalcy has more ethical implications. As Steven Rose pointed out during the Amsterdam meeting, the question then becomes one of the extent to which people should be adjusted towards the norm if they are diverging from it - be it for moral, commercial or political reasons.

As a matter of fact, many organic functions have a logical, physiological optimum. But how can one define such an optimum for cognitive, behavioural or emotional functioning? There is no definition of the optimum for reasoning or for a mathematical sense or for fear. Then, what should be considered to be the optimum for memory? Knowing that visuospatial memory reaches its maximum around the age of..., should it be considered that all individuals are below the norm after this age for this particular form of memory? And should age-related memory impairment be considered to be a normal consequence of ageing or to be a pathological condition? Similarly, one may wonder where the borderline is between moroseness and mild depression. Or, when a child 'misbehaves' in school, whether he falls into a medical category of deviant behaviour.

Moreover, what may count as 'normal' in a society changes over time. Thirty years ago, people with bad teeth were considered quite normal, whereas now, in most societies, they are frowned upon because 'you can do something about it'. Today, we expect our children to be intelligent, we train them in many ways to perform well on cognitive tests, and by doing so we up the norm of what counts as acceptable intelligence, which may result in social pressure. The field of what counts as 'normal' will then inevitably become narrower.

The continuum between normalcy and pathology can also vary over time for a given individual. Within the growing burden of brain conditions that affects or will affect the population in the coming years, many, among the most common ones, have a very progressive onset. Advances in the brain sciences allow a better understanding of the mechanisms underlying these conditions, and this leads not only to more effective treatment, but also to earlier diagnosis. In other words, the 'abnormal' state is slowly overlapping normalcy. That's where new questions arise: when does one become ill?; when should one begin to be treated? and to what extent?

Early diagnosis

All diagnostic techniques are nowadays achieving an unprecedented level of refinement, be it as the result of different techniques of brain-imaging, genetic testing for a predisposition to various troubles, biochemical markers identified with the help of neuropharmacology, neuro-psychological testing, or even on the basis of the more and more detailed DSM-IV for the establishment of statistical criteria of psychiatric disorders.

This is a very positive trend for the general improvement of public health, which often relies on early diagnosis and/or early treatment for many diseases. People could start changing their lifestyles or taking preventive medication long before the onset of a predicted pathological condition. This could result in much less serious symptoms or even complete avoidance of some diseases. On the basis of genetic profiles, it could even become possible to prescribe with more specificity safer and more effective drugs for particular individual profiles.

But starting a treatment, particularly in the absence of obvious symptoms, is a decision which has ethical implications that go far beyond the first logical idea of prevention. For example, should one give neuroleptics to prevent the onset of schizophrenia in a youngster with a known genetic predisposition? Or with typically perturbed brain patterns but no symptoms? Or even with 'only' weird behaviour? One knows the psychosocial burden of such a 'psychiatrisation' and its influence on the future self-identity of a young person. On the other hand, it is also well established that the later the treatment of real schizophrenia starts, the greater the psychosocial and cognitive handicap that develops. This is currently a point of much debate among psychiatrists, and the progress expected in diagnostic techniques is not necessarily going to help solve the question.

Another example. New batteries of tests, both genetic and neuropsychological, allow very early detection of Alzheimer's disease with growing accuracy. Besides, an increasing understanding of the mechanisms of memorisation have made it possible to identify various substances active upon the receptors involved in these processes, and some newly discovered compounds seem to be able to postpone the first significant manifestations of dementia. This is good news for many sufferers, and nobody would deny it. But still, questions arise. When should one begin to give these drugs to ageing persons showing a mild cognitive decline and slight memory loss? For how long? Who is going to pay? What about genetically predisposed persons: should they be given lifelong medication?
The trend is towards making a 'therapeutic bet' on the patient before he gets overtly symptomatic; and this trend is subject to numerous large-scale clinical trials which have not yielded clear results yet, but which are considered by some critics to be a manoeuvre of the pharmaceutical industry, playing on the natural anxiety of the population.

New diagnostic techniques make it possible to make predictions not only about future diseases but also about a variety of mental states. New brain-imaging techniques promise to feature personality profiles showing tendency to depression, behavioural tendencies like alcoholism or paedophilia, and personality traits like aggressiveness, antisocial behaviour or risk aversion. Even racial attitudes, political opinions and religiosity are said to be readable in the waves of the EEG! Consequently, different kinds of screenings might be proposed to the public in the near future, perhaps even without medical advice and on purely commercial grounds. One hardly dares to imagine what might be the societal consequences of scanning children for antisocial behaviour or for musical ability ...(see the section on 'privacy and personal freedom')

Are we medicalising society?
Another new issue in diagnostics is their extension to 'new pathologies'. For example, these last years have seen a remarkable increase in the number of psychiatric disorders. This is partially due to a refined diagnosis based on better knowledge of brain functioning. But numerous conditions that were considered within the limits of normalcy have also been turned into pathological ones, not only on scientific grounds, but also under some kind of commercial and/or social pressure.

The psychic field is characterised by an almost total lack of objective markers for defining disorders meaning that most diagnoses are based on consensus. The world-authoritative US-based Diagnostic and Statistical Manual (DSM-IV) was created in 1952 to harmonise the criteria of diagnosis in the far too imprecise field of psychiatry, and the standardisation of diagnostics that has followed has undoubtedly been beneficial both to patients and to the science of psychiatry. But according to numerous practitioners, it has now reached a level of systematisation that incites them to transform symptoms into diseases, which automatically brings up the notion of treatment, followed by the ad hoc medication provided by the pharmaceutical industry. As Steven Rose puts it, *"the offer to treat, to manipulate, to control"*[3].

It is also a well-described social phenomenon that any medical label stuck on daily life problems gives those problems some kind of acknowledgement - and exoneration? Moreover, it usually holds the implicit promise of a cure - sometimes against all rationality. We are currently experiencing a broadening of numerous medical concepts with, as a consequence of this, a decrease in their clinical accuracy. They become 'folk psychology' and are stretched to include ever more benign problems without any real scientific justification. The Dutch educationalist, Micha De Winter, has developed the example of ADHD children to illustrate this idea, affirming that many parents long for a medical label for their child, even if it is clinically doubtful that the child is pathologically different, because this label gives them a claim to special measures, and dismisses other explanations of the child's behaviour which would put the blame partly on the parents.[4] The French philosopher, Pascal Bruckner, for his part, has developed the idea of 'victimisation' that goes together with the growing medicalisation of society[5]. The pharmaceutical industry and patients' organisations are playing a crucial and very ambiguous role in this trend:[6] patients quite naturally require from governments the latest therapeutic treatments for their diseases, which serves the interests of the industry, which in turn sustains the actions of patients' organisations...

"Is something changing in reality which is actually increasing the psychiatric burden that we see at the moment? Or are we as a society becoming more aware of the problems which have always existed, and therefore do we see the need to treat and extend the range of the use of drugs?" asked Steven Rose during the Amsterdam workshop. Consumerist Western societies are said to have lost the sense of community that is woven in more traditional ones. Impersonal and individualistic ways of life prevent people from seeking advice and consolation in the family circle. Said Peter Molnar: *"They don't have the social capital, but they have the money. So people have to get help from unknown third parties, which they have to pay. And pills are definitely cheaper than interpersonal contact."* Are we treating with drugs diseases that are created by changing cultural values?

Steven Rose also pointed out that this evolution runs in parallel with a strong tendency to put the emphasis more on individual problems than on the social context, while Andreas Roepstorff remarked that *"everything that had a social component to it in the 1970's now has a neural component."* Trying to find the roots of aggression in the genes or in the brain pat-

---

3 'Brave new brain', article by Steven Rose, in the *Guardian*, May 8, 2003.

4 Micha De Winter, in *Een Ander Ik*, ed. by Rathenau Instituut, 2004, chapter 5.

5 Pascal Bruckner, *La Tentation de l'Innocence*, Grasset, 1995.

6 See for an elaboration on this role for example Dutch ethicist Tsjalling Swierstra in *Een Ander Ik*, ed. by Rathenau Instituut, 2004 chapter 6.

terns of young suburbanites is easier than concentrating on social solutions for this worrying problem. And giving drugs to calm down hyperactive kids in school avoids questioning the relationship between the child and its social and educational environment. Steven Rose: *"If we can no longer produce a more beautiful world by changing the one around us, not even in wealthy western Europe, then the choice is to say: 'OK, we go into individual solutions and look inside the individual and personal lifestyle choices', which easily leads to genetic explanations of problems'"*. Many participants supported the idea that the increasing tendency to assume that the right way to treat a problem is to find a drug to manipulate the brain is a tremendously important and worrying social issue. And Peter Molnar added that it engages our responsibility towards our children not to teach them that every problem has a pill-solution for it.

Enhancement

Normalcy has two frontiers: one in relation to the pathological field, which is the shifting line of medicalisation mentioned above, and one in relation to 'supernormalcy', and that's where enhancement begins. Treatment can be seen as bringing individuals back to species-typicality, whereas enhancement is improving 'normal' individuals beyond species-typicality.

"Do not adjust your mind. The fault is in reality" (anonymous graffito)

Towards superhumans?

Today, enhancement is achieved mainly by pharmacological means: memory enhancers, psychostimulants, antidepressants. In the coming years, increasing knowledge of molecular pharmacology of the brain will yield substances with a more specific target, which in turn will probably lead to fewer side effects, hence making them available on the free market and letting them be subject to free advertisement. This is the quite logical way things go, and as Ian Ragan, from Eli Lilly, puts it: *"My job and that of my industry is to produce drugs that are more safe and more effective than the previous generation. The safer they are, and the more effective, the more likely they are going to be used recreationally, to enhance normal behaviour."*

Memory enhancers were primarily thought to be dedicated to the treatment of Alzheimer's disease, a research field pushed by the majors of the pharmaceutical industry, which has already yielded some interesting compounds. But many small start-ups (with huge media coverage and tremendously growing shares) are also competing in this field with the overt aim of finding drugs that will specifically target pure memory enhancement, without any therapeutic vocation. A fabulous market potential. US sociologist and ethicist Paul Wolpe thinks that we will see those drugs on the market within a decade, but that they will begin to percolate into society substantially before that.[7]

7 James Butcher, 'Cognitive enhancement raises ethical concerns', in *The Lancet*, vol362, July 12, 2003.

Psychostimulants, for their part, already exist and are being marketed - though on prescription. Some of them are clearly improving performance as regards attention and executive functions (the higher cognitive functions that make it possible to plan and solve problems) and are thus used for enhancement purposes. Methylphenidate (trade name Ritalin™) is the most common one; it is officially marketed for children with attention deficit and hyperactvity disorders (ADHD) but its use largely exceeds this very specific indication. According to Martha Farah of the University of Pennsylvania, many university students are using it to enhance their ability to focus and study. And besides the obvious overuse of the drug to control unruly children in poor urban school districts in the USA, she also sees evidence that many wealthy parents are now choosing to give the drug to their well- behaved children to enhance their classroom attention. *"Will the day come when students are told to hand in a urine sample to prove that their achievements were the result of hard work and not pharmacology?"* asks James Butcher in a Lancet neurology editorial[8].

In the class of psychostimulants, there should also be mention of modafinil, a vigilance drug making it possible to stay awake for long periods of time, apparently without the side effects of amphetamines. Modafinil is officially licensed for a rare condition called narcolepsy, but it can also be bought on the Internet and is the object of much interest in military departments. It has been said that soldiers in Iraq have been on the drug during operations in 2003.

Antidepressants seem to be undergoing the same expansion. The new generation of these drugs (generally called SSRIs, the best known of them being Prozac™) is much safer than the previous generation producing fewer side effects. This has allowed a much wider use of them, firstly within the medical field, where applications have been rapidly expanding: milder and milder forms of depression, social anxiety disorders, obsessive compulsive disorders, etc. But another trend is the use of these substances in a more recreational way, as 'mood-brighteners', by people considered to be perfectly healthy. In the report of the US President's Council, Leon Kass observes that *"people who take SSRIs in the absence of definite mental illness commonly report that negative feelings such as sadness and anger do not disappear but diminish, as does the inclination to brood over them. Loss, disappointment and rejection still sting, but not as much or as long (...) self-esteem and confidence rise. One is more easily able to experience pleasure and accept risk. Mental agility, energy, sleep, and appetite become more regular, typically increasing. And mood brightens- though not to the point of perpetual bliss or anywhere near it".*[9] In other words, the small worries of everyone's daily life fade away, and people definitely see the bright side of life. Almost the perfect 'soma'?

8 James Butcher, 'Cognitive enhancement raises ethical concerns', in *The Lancet*, vol362, July 12, 2003.

9 *Beyond Therapy, Biotechnology and the Pursuit of Happiness,* A report of the President's Council on Bioethics, October 2003. www.bioethics.gov.

Non-pharmaceutical means of enhancement and/or treatment are still at a very rudimentary stage, but deserve to be mentioned in a report like this. The Transcranial Magnetic Stimulation (TMS) technique, which has so far been mainly a research tool, was actually tested for the relief of depression, post-traumatic stress disorder, schizophrenia etc. It has also proved capable of affecting the cognitive functions, while causing no pain to the subject. Although nobody is quite sure how it works, it has been suggested that TMS can improve performance in memory and reasoning tasks, rouse people from the effects of severe fatigue, or teach them a new skill,...all very promising arguments for the emerging market of 'neurotechnology': *"While the first commercial TMS systems have been large, power-consuming units more suitable for clinical environments than home use, it is not out of the question that more portable or even wearable systems could emerge down the road. If so, the neurotechnology industry could have a powerful and profitable new weapon in its arsenal, in addition to a reduced threat of invasion to the patient".*[10]

In the long run, more SF-like techniques will probably be developed. One can mention here research in the fields of artificial intelligence and brain-machine interfaces developed to augment memory and machines controlled by thought only.

The easy way?

Is it ethical to enhance one's own functioning? Is taking a stimulant before an exam any different from gulping down a cup of coffee? As Paul Wolpe declared at the first Neuroethics conference in San Francisco (2002), *"we have a hard time accepting a big gain with no pain because we believe hard work ennobles us".*[11] Perhaps we are being inconsistent and too puritan in avoiding these biochemical methods, but on the other hand, the process of reaching a goal 'the hard way' may provide collateral benefits as one learns something about oneself along the road, which the short-circuit would not. As a matter of fact, ethicists worry that, if cognitive enhancers are massively used, it might change human society and the values it cherishes, leaving behind the merits of effort in favour of shortcuts to better performance. Just as in sport....

The problem is slightly different with the so-called 'mood-brighteners' like Prozac™, because of one additional dimension: they are used firstly to alleviate some form of suffering. But *"how much human suffering can we tolerate before prescribing a drug?"* asked psychiatrist Johan Den Boer, in one of the debates of the Amsterdam workshop, *"we all suffer a little bit sometimes, but we don't necessarily take drugs for it."* Is it so that suffering - as well as joy - is a part of life that we cannot and should not escape?

---

10 Seen on www.neurotechreports.com, a website which also claims that *"Neurotechnology, the application of electronics and engineering to the human nervous system, has now reached a level of commercial and scientific maturity that will produce enormous benefits to patients and profits to investors."*

11 *Neuroethics: Mapping the field.* Conference Proceedings, May 2002, San Francisco. www.dana.org.

The answer of the representatives of patients' organisations and of Jes Olesen, President of the European Brain Council, was straight: *"People who don't suffer tend to think that suffering is an important part of life, but people who suffer want to be treated! It is important that normal persons don't impose limits on those who are suffering because they are not in a situation to understand what it means."*

Moreover, patients' organisations claim autonomy of the patient and respect for his needs: *"The need starts in the perception of the patient, not at some point of suffering. It may have no relevance to what the scientists expect, and if the scientists are there to provide help for people, then they should try to provide help at this point of need"*, said Alistair Newton.

As a matter of fact, it is nowadays commonly admitted that the relief of physical pain should be immediate and even preventive. But it remains unclear whether the relief of psychological pain 'should' also be achieved by pharmaceutical means. According to Johan den Boer, *"the great danger is that it might induce a regression to the mean, by preventing people from experiencing a range of conditions, life situations etc. - not necessarily suffering - but that will make us all more similar to one other. Lack of diversity in psychological states is a danger to society as a whole in the long run. It brings up Aldous Huxley's visions of the future in a worrying way."*

In the same set of ideas, Susan Greenfield worried that a drug-dependency culture could be imposed upon the young generation, with a premium being put on just living for the moment, 'just chilling out': *"I fear a future where people become passive, where they are brought up in an environment where they haven't had to struggle over some of life's problems, where as soon as you feel a little bit unhappy you take a drug and where you no longer have to think about the significance of what happens because everything is in the moment."*

Some philosophical arguments are also questioning the fact that enhancement is defying the 'natural' limits of humanity. Some, like Francis Fukuyama, author of the best-seller *Our Post-Human Future*, sees in the advances of science the risk of losing the essence of being human - though he doesn't clearly define what he means by this 'human essence'.

Trans-humanists, for their part, dream of taking over control of human evolution in order to redirect it towards supra-human individuals. Between those two extremes, many argue that we are already post-human, since we wear eyeglasses, use insulin, have artificial hips and ride on airplanes and that there is no 'natural' boundary beyond which we would not be humans anymore. Another argument is that it is impossible to consider humans independently from the culture and environment that they have shaped and that have shaped them in return.

But finally, as James Butcher stated in his issues for the Amsterdam meeting, perhaps we should not be worried so much about a 'post-human future', but rather about a future in which we recognise ourselves as humans but others do not perceive us to be the same person we once were. There is actually fear that enhancers might disturb individual perception of the self. If their use becomes commonplace, it could lead to some loss of authenticity and to some new forms of identity troubles. The central question would then be: does this pill estrange me from myself? Or do I become more myself? Is it a drug-induced personality or has the drug cleared away the barriers to my real personality? As it happens, the brain sciences almost inevitably bring up philosophical questions like this one of personal identity, which will be elaborated upon in the next section.

Personal Identity

There are not many interesting ways in which the liver of a healthy person differs from the liver of another healthy person. That is why the idea of a liver-transplant typically does not meet much metaphysical resistance. Whereas the whole idea that one's brain could be transplanted is simply inconceivable - not only because of the absolutely insurmountable practical problems one would face, but also for philosophical reasons. As William Saffire, chairman of the Dana Foundation, puts it, our brain is our 'organ of individuality'[12]. So, to 'change one's brain' is to lose one's self as we know it.

This, in fact, opens up the question of 'personal identity': what is it to be the same person today as one was in the past or will be in the future? By now, it is possible to implant cochlear devices in the inner ear or install electrodes deeper in the brain to diminish the symptoms of Parkinson's disease or Tourette's syndrome. These techniques will surely be developed and extended to other afflictions. To many people, it is an eerie idea that inorganic material functions in the brain. It makes Flavio Keller wonder if there is *"an ethical necessity to put some limits to the physical manipulation of the brain to safeguard personal identity"*. And sociologist Paul Wolpe states: *"Once you integrate technologies into the brain, you have to ask yourself the question, is there an end to the non-technological me and the beginning of the technological me, or is it now all me? Am I part technology and part organic?"*[13] The answer to this last question, at least, seems to be 'yes'; we are already cyborgs: part organism, part machine. And this extends to our brains as well.

Changing our material

This situation is certainly unfamiliar, but perhaps not as threatening to our sense of self as one might think at first sight, depending on one's definition of the self. For philosopher Patricia Churchland the 'self' is a *"set of capacities that involve not only representations of the body itself, but also representations of internal aspects of the brain - the brain's mental life."*[14] So, we certainly need to change our representation of our body if there are electronic devices implanted in our brain, but this in itself need not force us to change our view of our identity. It seems much more important that psychological relationships such as memory and character are preserved in the process. Thus you are still you if your mental life is intact - regardless of whether organic material (like cellular grafts of stem cells) or inorganic material (like electronic implants) helps to sustain this mental life.

Changing our experiences

This is not the end of the challenge which the brain sciences offer to our

---

12 *Neuroethics: Mapping the field*, p. 7.

13 *Neuroethics: Mapping the field*, p. 164.

14 *Neuroethics: Mapping the field*, p. 20.

concept of personal identity though, since they do offer many techniques for changing our memories and character and these techniques are already used on a grand scale. SSRI antidepressants like Prozac™ and Seroxat™ chemically change our moods over a long period of time, thereby arguably changing our 'character'. As mentioned at the end of the last paragraph, one of the central questions relating to mood and cognition enhancers is: does this pill estrange me from my true self? Or, on the contrary, does it enable me to become my true self? Will I cloud my view of who I am by taking this drug, or will the drug clear away the barriers to my real personality? Answers to this question have not yet crystallised completely and will probably differ according to individual situations.

Furthermore, at this moment a lot of research is going on into devising psychotropes that enhance or erase memories. Here questions of personal identity also come to the fore, since, as Richard Morris stated during the workshop, *"memory helps to define our individuality"*.

The social self

Interestingly, the rather insular notion of self which in European culture is connected to personal identity might need to be altered in view of findings in the brain sciences. As biological psychiatrist Johan den Boer points out: *"A brain simply does not develop when it is not situated"*. We are not born with a personal identity, we become persons only by interaction with other persons in our environment. Many of the current frontiers and thresholds in cognitive neurosciences lay in establishing the social networks which the brain needs to make sense of its world and itself. During the workshop, psychiatrist Roland Jouvent stressed the recent discovery of 'mirror neurones': similar brain areas are activated while performing an action oneself as while observing the same action performed by somebody else. The same holds true for emotions: we seem to understand the emotion of a fellow human by simulating the emotion within ourselves - a biological demonstration of the cultural notion 'empathy'.

For Jouvent, seeing others as similar to ourselves is a general feature of human cognition. In fact, findings by Jean Decety in Seattle seem to indicate that we need a specific faculty in the brain to make us realise that the actions or emotions which we experience are (or are not) our own. A defect in this so called 'agentivity' faculty might correlate with inflictions such as schizophrenia. This exciting new research points out that 'personal identity' is a concept which we probably need to think through again. Our sense of self might be a sophisticated, fragile asset, necessarily based on our engagement in social relations.

The new definition of death... and of life

Finally, what becomes of our identity when death takes place? Until recently, death was established when the heart stopped beating. But the development of reanimation techniques has made it possible to keep hearts beating and lungs breathing artificially, which has led to a need for a clear definition

of death amongst clinicians. In the 1970s, it was almost unanimously decided that death would be the complete cessation of any brain activity, established by flat EEG and/or confirmation of the absence of blood flow within the brain (made possible by various imaging techniques).
In the last years, though, imaging techniques have made possible much more subtle distinctions in the functional assessment of the brain. For example, it has now become possible to distinguish between different depths of comas and to establish rather reliable prognoses for their outcome. Hence it also becomes possible to assess which part of the brain is 'dead' or still 'alive'. In other words, from now on, there could be several ways to be dead!

Then there comes the inevitable question: what is human life? What is a person? Is a person still human when there is no hope that he/she will ever recover any form of consciousness (linked to the function of the cortex) - even if automatic reflexes (linked to the function of the brain stem) are maintained? But then, one little step further, one could consider that the loss of the functions that make the identity of the person would be equivalent to the death of this person e.g., in the advanced stages of Alzheimer's disease. Some scientists and philosophers already worry that, if no reflection takes place on this crucial question, we could see a shift towards abusive extension of the definition of death for economic reasons: "*the growing financial and psychological burden on the living and the increase in neurodegenerative diseases due to an ageing population will push us in that direction despite strong protests*". (Paul Churchland)[15].[16]

---

15 *The Engine of Reason, the Seat of the Soul: A Philosophical Journey into the Brain*, by Paul Churchland, 1995, 307.

16 For an elaboration on this, see *Brain Policy, How the new Neuroscience will change our Lives and our Politics*, by Robert H. Blank, Georgetown University Press, pp 36-43.

Reductionism and clashing vocabularies

A self cannot exist without a brain. That is now a fact of science which none of the workshop participants wanted to deny. On an ontological level, mind and brain are the same: for everything that happens in the mind, there must be a neurological correlate. But, on an epistemic level - that is, as a knowledge-system for understanding ourselves - mind and brain open up very different practices and experiences of ourselves.

This is relevant for European citizens, because 'brain-talk' is no longer just a tool for understanding and curing the brain; it is on the verge of becoming a cultural narrative. In the words of Andreas Roepstorff: *"Neurons and the brain are moving out of the hands of the brain scientists and into becoming a kind of important metaphor for thinking about what it is like to be human."*

Mind = brain = molecules

The crudest proponents of this new 'cosmology of the brain' reduce our mental life to the movement of molecules. These reductionists *"seek to integrate the mind and mental phenomena - fear, pain, anger and the like - with the natural world by showing them to be natural phenomena. Their inspirations are the famous reductions of science: of the heat of gasses to molecular action, et cetera. Reductionists hope to show a similar relationship between mental kinds and neurophysiological kinds."*[17]

Ever since the European Enlightenment, reductionism has been a guiding idea of science. *"What you want to do in science is look for reductions"*, states philosopher Patricia Churchland[18]. Ian Ragan of the pharmaceutical company Eli Lilly supported this idea during the Amsterdam workshop: *"In trying to come forward with treatments, we do have to use a reductionist approach, rightly or wrongly."* Pharmaceutical companies achieve impressive results by taking this approach, as is made clear by the popularity of psychotropes like Prozac™ and Ritalin™. But this in itself does not prove that the 'reductionist' model is right when applied to the brain. *"The technologies may work, even if they are based on bad models"*, Steven Rose pointed out.

In a situation where 'brain talk' is starting to 'colonise'[19] our way of understanding ourselves, we need to think carefully about the justifications for reducing the mind to the brain. Is reducing the mind to the brain really comparable with reducing heat to the motion of molecules? Axel Cleeremans states: *"The mind definitely is the brain, in the specific sense that there can be no mental event that is not also a neural*

17 Jaegwon Kim in *Routledge Concise Encyclopedia of Philosophy*.
18 *Neuroethics: Mapping the field*, p. 55.
19 A phrase of Thomas Metzinger.

event." But it does not follow from this fact alone that the level of the mind can be reduced to the level of the brain. That is, this in itself does not prove that the brain is fundamental to the mind.[20] So, even now that we are progressively able to link Alzheimer's disease, loss of memory, or psychiatric syndromes to neurological patterns, we should keep in mind that we are in fact not doing much more than just that: linking different sorts of observations to each other. If one assumes that this link implies that the neurological pattern is the cause of the symptom, than, technically speaking, a correlation is mistaken for a causality.

Clashing vocabularies
As has been said before, a human brain will develop only within a social setting and it retains its plasticity for a long time during its life. As Axel Cleeremans puts it: *"We find traces of the brain's constant interaction with its environment and with the culture in which it is embedded in the very structures that support our mental life."* So, if we consider the human brain to be an assembly of neurons, we will be right. But we will also be right if we consider the human brain to be 'the minutes of our experiences'. In fact, it is up to us to choose the way we want to look at it. Detlef Linke pointed this out when he remarked during the Amsterdam workshop: *"Epistemologically we are in a situation where we have to ask ourselves which complexities to choose."*

This choice is not without consequences - in fact, it will form our practices. For example, by framing depression in a biochemical or neurological way, the cure or alleviation of the affliction is logically sought in the same direction. By choosing this 'episteme' a treatment comes to the fore which specifically addresses the brain, like psychotropic drugs or even neurosurgery. Other ways of alleviating suffering, for example training mental skills or changing the social surroundings of the afflicted person, do not come into focus.

This even led Steven Rose to explain the current popularity of reductionistic explanations partly as a backlash against the social actions of the 1970's - and in comfortable concordance with the dominant liberal outlook in Europe. Should these reductionistic explanations take over our reasoning, being sick or unhappy would no longer be the fault of society - or oneself. It would be the fault of one's brain. So, nobody would be responsible and nothing need be done, except looking for pills or other means to cure the brain (see the section on 'medicalisation' for an elaboration on this).

Mental diseases are at the cutting edge of this discussion on which vocabulary to choose (in fact, the issue at stake drives the terminology: when exactly do we call a disease mental?). Of course, the practice of attributing responsibility for mental diseases like schizophrenia, depres-

---

20 See Johan den Boer, *Neurofilosofie*, for an explanation of our tendency to assume that, since the mind is the brain, the brain must be fundamental, pp. 32-35. On the first day of the workshop, this problem popped up during the talk of Mark van Buchem. There it was pointed out that, although the tendency in biology is to suppose that structure precedes function, this is probably not true for the brain. There it might well be 'a two-way affair', that is, brain function might as well, in some cases, precede brain structure.

sion or anorexia nervosa to (the family of) a person who suffers from it can be very cruel, as Paolo Lucio Morselli pointed out during the Amsterdam workshop. On the other hand, it might be a very real loss of insight and experience for our culture if the knowledge that psychiatrists and psychotherapists bring to this field is downgraded because of our enthusiasm for brain talk.

Inevitable psychology

Some people fear that the hype surrounding 'hard' neurosciences might gobble up the 'softer' knowledge of psychologists and psychiatrists. But the truth is that psychological insights are necessary even within a neuroscience framework, because we cannot understand the workings of the brain without 'mental' concepts. Even at a practical level, speaking about 'electric activity' or a 'discharge of neurotransmitters' is often not informative enough. In order to say what this activity means, neuroscientists have to rely on psychological concepts.[21] And psychological terms are the obvious heuristics for cognitive neuroscientists setting up their research models. The current tendency is for the neuroscientist and behavioural scientist to work together in a real attempt to integrate vocabularies. For example, the coming Diagnostic and Statistical Manual of Mental Disorders (DSM-V, to be published around 2010) will explicitly link 'traditional' psychiatric patterns to the pathophysiology of the brain.

So, brain studies are typically a field in which different vocabularies meet: physics-oriented science on the one hand and knowledge in the form of systematised psychiatric and therapeutic experience on the other. Those vocabularies have to be joined at some points as brain science matures. The exciting and challenging question is: how exactly? It remains doubtful if this will entail a pure reduction of mind to brain, because, in the words of Axel Cleeremans, in our mind/brain *"there might emerge something at some level-plus-one that you cannot reduce to a simple one-to-one correspondence with what is happening at the lower level."*

In the meantime, one need not be too afraid that our social practices will collapse if our mental life and moral values are 'reduced' to the motion of molecules. We might keep in mind the consoling words of Antonio Damasio: *"The fact that acting according to an ethical principle requires the participation of simple circuitry in the brain core does not cheapen the ethical principle"*.[22] And as Andreas Roepstorff remarked during the workshop: *"What if the mind equals the brain? You can take the perspective that all the phenomena that we know exist in the human mind somehow have to exist in the human brain: in other words, rather than reducing the mind to the brain, the idea is that what we know is taking place in humans has to take place in brains."* So, according to him, good neuroscience should explain how what we know to be happening in our minds is also taking place in our brain. And not explain these phenomena away.

---

21 This point is made by the psychiatrist Willem van Tilburg in *Een ander Ik*, ed. Rathenau Instituut, 2004, chapter 4.

22 In *Neuroethics: Mapping the Field.*

Responsibility and Free Will

The last paragraph spoke about the tendency to reductionism manifest in the work of some neuroscientists. By that we meant the desire to 'reduce' the level of mental phenomena to the level of physical, material ones. This paragraph treats the related, but logically different, concept of determinism. By 'determinism' we mean the view that *"everything that happens is necessitated by what has already gone before, in such a way that nothing can happen otherwise than it does"*[23], human actions not excepted. So, where 'reductionism' talks about *levels*, 'determinism' takes *time* as its axis.

Determinism thus understood seems to pose a problem for our concept of free will, for it means that we are not free to do as we like, but are caused ('determined') to act as we do. This classical problem is not created by the brain sciences but is made especially relevant by them, because we typically hold that our 'free' decisions take place somewhere in the mind/brain.

According to our deep-held beliefs, we are responsible persons who can freely decide to act and therefore are accountable for our deeds. Of course, there are exceptions to this rule; we do not hold little children or people with serious mental disorders responsible for their actions. We also know from personal experience that a lot of our deeds are not thoroughly thought over. Still, normally we assume people to have free will when there is no neurological or psychological dysfunction or immaturity discernible, and we have good reason to do so, since there are enormous political, ethical, and religious implications attached to this image of ourselves as free and responsible humans. Without this assumption of our freedom, our complete legal system seems to lose its foundation and much of our art becomes unintelligible.

The classical scientific model

Unfortunately, concepts like responsibility and free will do not fit easily within a scientific framework. In the scientific paradigm, a certain cause is taken to be necessarily followed by a certain effect. The challenge scientists see for themselves is to find out the rule behind this pattern. If you succeed in formulating this rule, you can in principle predict what will happen, given a certain 'cause' or input, and provided that there are no other variables at work. So ideally, scientists formulate rules which will predict the behaviour of physical objects, including human beings qua physical objects.

Of course, almost every brain scientist will admit that it will not ever be possible to predict all human behaviour: too many factors, known and as yet unknown, influence an organism with our degree of complexity. But

---

23 *Routledge Concise Encyclopedia of Philosophy.*

this does not hinder research programmes in the brain sciences from working from the scientific set of ideas and presumptions that all events must be *caused*.[24] For example, the process of decision-making is studied in terms of nerve connections and electrochemical reactions where the output of such interactions is conceived to be determined by the input - even if in practice it is impossible for the researcher to take all these interactions into account. The ultimate consequence is that people may have the idea that they are *doing* something, whereas in the eyes of a 'hard' determinist there is just something *happening* inside them.[25]

According to this view, we are not the authors of our actions. Since this bold scientific view is obviously at odds with the religious or metaphysical beliefs to which most people are dearly attached, it comes as no surprise that they feel very threatened by it. That is why neuroscientists have to face *"fears in the mind of the public that if we say something like 'you are just your neurons', free will and personal identity will evaporate, degrading a human being to just a bunch of neurons floating around, that is: to zombies"*, as Axel Cleeremans puts it. And according to Detlef Linke, the deepest fear of citizens is that the concept of freedom might be endangered by saying: 'you are your brain'. This fear is logical, he states, because *"freedom was formerly defined as being against natural processes."* That is, the only sense in which to understand freedom is decisively at odds with the determinism prevalent in the study of natural processes.

Dualism versus materialism

Thus, none of our options seem attractive. We either postulate two dualistic 'spheres', one spiritual and one physical, and save human freedom by postulating that the spiritual sphere is free from the laws of physics - thereby importing the huge problem of how to conceive of the interaction between these two spheres. This was the solution, and the problem, of René Descartes and his followers. Or we 'reduce' every phenomenon to the physical world, thereby threatening our human moral values and our authorship of thoughts and actions, which is the challenge determinists have to take up. The choice seems to be between viewing ourselves as (physical) zombies or (spiritual) ghosts.

As said before, brain scientists have to think about reconciling these two domains - if only from a public-relations point of view. The question then becomes, taking up the cue of Detlef Linke: how can we redefine freedom so that it is grounded in natural ('caused') processes?
The answer to this question is still heavily debated in philosophy and cognitive neurosciences, but progress seems to be made by using a more refined vocabulary. Biological psychiatrist Johan den Boer says: *"If you*

24 The words of Patricia Churchland might be helpful here: "Predictability and causality are not the same. Some caused events are not predictable." *Neuroethics: Mapping the Field*, p. 51.

25 'Determinism' comes in flavours; for a fast overview see, for example, the dilemma of 'free will and determinism' in A.R. Lacey's *Dictionary of Philosophy*.

look at the [neuroscientific] work on decision-making, you often get the impression that parts of the brain are involved in reward, in emotional processing, in cognition, and in decision-making." According to him, it is a fallacy to attribute *"psychological capabilities which only human beings as a whole can perform to isolated parts of the brain."* Den Boer says: *"I am the one who is taking the moral decisions, not a part of my medial prefrontal cortex. [The medial prefrontal cortex] may mediate the proper functioning which in the end leads to me, as a whole person, taking the decision. But that is completely different from saying that the medial prefrontal cortex takes the decisions. That's wrong."*

For this reason, neurologist Antonio Damasio warned at the Dana Conference in San Francisco against talking about 'centres' in the brain for this or that faculty; this picture seduces us into thinking that a decision or action is produced in that centre.[26] Like many other modern brain scientists, he prefers to talk in terms of brain systems. And these systems can be very broadly defined, even to the extent that the social environment is part of it.

Compatibilist views

So, it will help if we do not configure ourselves as the audience to decisions being made somewhere in a part of our brain. We are involved in our decisions. We are part of the network and thereby part of the ways in which the past determines the future. It is probably there that we can pinpoint our responsibility, although it is obvious that a lot of conceptual and scientific work has to be done before we can talk about this with more sophistication.

One important thing to remember: we may not be conscious of the ways in which our brain and body determine our thoughts and actions. But this cannot lead us to conclude that we 'know' that our thoughts and actions are not determined. Since our own subjective observations of our inner life do not offer anything decisive on this topic of determinism, we need not see the reconciliation of the subjective and scientific view as an insurmountable problem.[27] Or, to state it differently, our problem would be far worse if our inner observations were to conflict with the scientific view!

Most of the more conceptually inclined researchers in the brain sciences hold that free will, if understood in a certain sense, is compatible with determinism, and are therefore called 'compatibilists'.[28] Obviously, discussions like this soon become highly technical and therefore may not be useful for the purpose of reassuring the general public. Still, it is good to keep the words of philosopher Patricia Churchland in mind: *"The fact of causality in the brain does not imply that there is no responsibility"*.[29]

---

26 *Neuroethics: Mapping the Field*, p. 17.

27 See Simon Blackburn's *Think: A Compelling Introduction to Philosophy* (chapter 3) for a development of this argument.

28 Eminent thinkers like Wolfgang Singer, David Chalmers, Jaegwon Kim and Thomas Metzinger are seen as compatibilists.

29 *Neuroethics: Mapping the Field*, p. 26.

Legal responsibility

This rather academic discussion becomes very relevant in a juridical context, and this is something citizens intuit perfectly. For our juridical system would be greatly undermined if its central concept, responsibility, was really proved to be redundant by the brain sciences. Traditionally, we hold that human freedom entails responsibility and that people who seriously misuse their freedom deserve punishment. But it would become almost impossible to convict any wrongdoer if one held that he or she was determined by the past to do what he or she did. How can one be guilty for an act which was not to be avoided?
In cases of juveniles or mental patients, courts sometimes already accept that a wrongdoer cannot be held responsible for his or her deeds, as was mentioned at the start of this paragraph. But this should not be declared too often, because in a way it parasitizes on the juridical system. As Johan den Boer pointed out: *"The fact that there is something wrong in the brain of serial killers does not lead automatically to the conclusion that they are not responsible for their deeds."* One might even say that the fact that you are known to have an inclination to misbehave in certain circumstances increases your responsibility to control yourself in those circumstances.[30]

It might seem humane and decent to try to understand and explain wrongdoers - to see them as patients instead of criminals. But, as the philosopher Peter Strawson mentioned, the flip side of this attitude is that a lot of what makes our human relationships *human* gets lost in the process. We start treating these people as being fully determined by their history and circumstances, which is a rather insulting thing to do. According to him, we should avoid doing so as long as possible for reasons of decency.[31]

The fact is that even though, for example, a mental disorder is caused by a complex pattern of electrical discharges in your brain, you will not experience yourself as someone with an electricity problem. This is just not the way we think about ourselves and experience ourselves. Given the recent excitement surrounding the brain sciences and the powerful position science holds in our society, we must take seriously the possibility that scientific concepts 'overrule' our metaphysical phenomenological concepts. This in itself is no problem; our concept of memory has also changed in concordance with scientific facts. In the same sense, our concepts of free will, responsibility and the like might have to be adapted too. On the other hand, brain scientists should be careful not to become unconscious proponents of a blunt determinism which leaves no room for freedom as properly understood, because *"otherwise there is no sense in which society can continue to function"*, as Axel Cleeremans warns.

30 Maureen Sie in *Een Ander Ik*, chapter 6.
31 Peter Strawson, 'Freedom and resentment' in Gary Watson (ed), *Free Will.*

Social justice

The burden of brain disease in Europe, as extracted from WHO's figures by Jes Olesen, President of the European Brain Council, is surprisingly high: 35% of the total burden of all diseases[32]. Consequently, EBC, through the voice of its President, asks for a more thoroughgoing teaching of brain science and brain diseases at medical and nursing schools, a greater proportion of research funds to be allocated to basic and clinical brain sciences, and sufficient efforts as regards prevention, primary care and hospital treatment for these patients. Jes Olesen also underlined that society will have to address the fact that the burden of brain diseases will further increase in the next 10-20 years due to the ageing population.

Besides this general effort, it is also of prime importance that every individual have access to the necessary treatment and preventive measures required by his or her health status. Assuming that this is more or less the case in most European countries (!) for basic healthcare, what will become of the newer and ever more expensive cutting-edge technologies? How far will national social security schemes be able to follow the exponential trend of health costs of brain diseases? Who will benefit from the newest brain-imaging diagnostic techniques, the early treatment of degenerative diseases, or targeted radiotherapy? Will deep brain stimulation interventions or other electronic devices be routinely reimbursed? What about gene therapy for brain tumours, stem cells grafts, neuroprotective drugs or growth factors, if they prove to be efficient? And above all, who should decide about this and by what criteria? It is already established in most countries that the choice of therapy is dependent upon the financial resources of the patient, but this two-tier medical care is by all the evidence particularly susceptible to divergences in the field of brain health.

The particular case of psychotherapeutic approaches has been stressed by some participants of the Amsterdam workshop. The concern is that, with the help of brain-imaging techniques, it will probably soon be possible to assess and validate the 'objective' outcomes of various forms of therapy. This could lead to the linking of the reimbursement of treatment to some 'objective' criteria without setting a price on the well-being and the subjective feelings of the patient. As Flavio Keller remarked: *"The uniqueness of subjective experience might not be easily grasped by this so-called objectivation of brain activity. Listening to the patient remains as important as ever."*

Cognitive enhancement and social justice

Now, moving from the field of health care towards the more debatable techniques of cognitive enhancement, questions of distributive justice almost inevitably arise, because of the potentiality of those practices to shake the very foundations of democratic societies. Assuming that, if it becomes possible to enhance brain functions when there is no medical

---

32 J.Olesen & M.Leonardi, "The Burden of Brain Diseases in Europe", *European Journal of Neurology*, 2003, 10:471-477.

necessity for doing so, this enhancement will probably not be funded with public money, there will thus be no equal access to these techniques within a community. Only wealthy people will be able to profit from this advantage which will make them even more competitive, and widen the gap within the society. Not to mention the abyss between the Western world and the rest of the planet.

Many ethicists plea for a proactive reflection on the societal impacts of cognitive enhancement because it will cause society to change incrementally, and, according to some, probably more rapidly than we expect. If the level of 'normal' cognitive performance rises, but only one part of the population has the means of attaining this new level of normal, it will reinforce the social pressure and set new societal norms. Take your pill and join the rat race! At a meeting of the New York Academies of Science, in June 2003, Judy Illes (Stanford Centre for Biomedical Ethics, Palo Alto) declared *"It is definitely a concern, as it will have a downstream impact on education, employment, and other opportunities for both children and adults, and it has the potential to change the very fabric of society"*[33]. But others are not so alarmist, pointing out that education is also overtly inequitably distributed, though nobody would imagine banning it for that reason. On the other hand, cognitive enhancers might be relatively easy to distribute widely. However, the threat might then come from another direction: widespread availability may itself cause population homogeneity, loss of diversity and commodification of human talent.[34] In Amsterdam, Axel Cleeremans took the example of what happens with plastic surgery to ask if we really want a society in which not only physical appearance but also mental states would have to correspond to some collective stereotypes.

Profound changes in society

At this point it might be useful to stress out that brain technology writ large would globally reduce the number of ways acceptable to be a person. If personality traits and behavioural characteristics become detectable from brain images and optionally treatable, people will be more inclined to discriminate against the bearer of those traits. This could lead to a less tolerant and very normative society where all deviations from the normal would be considered as deviant and 'to be treated or re-educated' (see 'privacy and personal freedom' section).

Another particular point of social justice, energetically raised by Zoltan Nemes at the Amsterdam workshop, is the issue of ageing societies in relation to cognitive enhancement and other forms of 'eternal youth medicine'. *"Possibilities to arrest ageing-related cognitive decline will create a gerontocratic and less competitive society. I foresee that the younger generations will be very frustrated by the fact that they can access only property, social in-*

---

33 Quoted by James Butcher in 'Cognitive enhancement raises ethical concerns', *The Lancet*, Vol. 362, July 12, 2003.

34 More about this? Don't miss the inspired SF interpretation by James Butcher in the Lancet: TetraTab, Cognitive enhancement gone wrong, in *The Lancet Neurology*, Vol. 2 , August 2003.

fluence and political power when the generation of gerontocrats will die out." Thus, what could indeed be beneficial for individuals (e.g. the prolonging of life) has to be considered to be a problem on the scale of society: *"The problem is not so much the prolonging of life, but the turnover of generations. Intergeneration conflicts have always existed, but brain science is worsening them by slowing down the natural turnover of power, wealth and values."*
Involving the citizens

The dilemma between individual benefits and societal values is probably the core of the question for all aspects of brain-based enhancement. Enhancement is neither right nor wrong, but is a paradigm of a technology that has to be discussed by the citizens to determine how it will be used. A series of questions has to be considered e.g. security of use, freedom of choice, use by children, prevention of coercion by employers or schools seeking higher-performing workers or students, etc. The right of not-enhancing has to be firmly established, though it is obvious that it will be difficult to set limits to the creeping incitement of marketing strategies, advertisements and social pressures. It is a matter of fact that the 'tidal wave of pills' increasingly proposed to the consumer could lead our Western societies towards some kind of Brave New World where we would all be bright, young, competitive and fit, never depressed, never aggressive, never tired... in a word, the perfect workers, perfect consumers, perfect citizens. Passive citizens never questioning the social and economic smooth running of the world.

"We could have this sanitised world where we would all be comfortable but we would have lost something" said Baroness Susan Greenfield, "but conversely, we could use all this new technological potential to gain insights into what it means to be an individual, to understand how the brain becomes unique and how we can preserve that individuality. How we can nurture in society a respect for individuality in other people and enjoy their diversity rather than trying to be better than the others."

The question remains open as to whether these decisions should be regulatory ones, or left to individual choice. Many participants of the Amsterdam workshop agreed that once a new technology is available and approved (or sometimes not even approved), people will use it, no matter what it has been approved for, unless very strong regulations frame its use. According to Wrye Sententia, it is looking more and more likely that the regulation of cognitive enhancement is going to be a prohibitive one in the US, which she thinks is a mistake. Quoting the President Council's recent report: *Beyond Therapy and the Pursuit of Happiness*, she deplores its manner of telling people what is the right form of the pursuit of life and what is not. *"We need to find other models to approach how to reconcile cognitive enhancing as a personal choice, but also at the same time bring the public along so that there is not decision-making in ignorance. I think society would greatly benefit from education on the complexity of these issues, rather than have a top- down injunction as to what is and what isn't acceptable."*

Privacy and personal freedom

Brain imaging is opening new and unexpected insights into individuals' brains and hence into their minds. This urges us to redefine the limits of privacy and personal freedom.

Recent advances in brain imaging and the computer sciences are yielding ever-increasing possibilities as regards obtaining precise and extemporaneous pictures of individuals' brains and a growing body of information about their actual health status. Since these data are already computerised, their storage and transmission are very easy. Their exploitation too. Brain-image databanks, like genetic databanks, can provide fantastic tools for medical research, and this should be kept in mind in order not to hinder the work of scientists. But one remembers also the controversy which arose from the "purchase" of the Icelandic health insurance databank by the genetic biotech company DeCODE. It is thus urgent to formulate guidelines for the correct use of this rich datafield.

The new brain-imaging techniques, combined with genetic or cognitive testing, are not only providing an opportunity for better and earlier diagnosis of medical conditions; they also pretend, in a way, to *predict the future* by detecting a proclivity to a widening range of diseases like depression, schizophrenia or Alzheimer's disease. They also promise to be able to define mental states and personality profiles such as tendencies to aggressiveness, to antisocial behaviour, to risk aversion and even to racial attitudes, alcoholism or paedophilia. All this has an enormous potential for 'hacking' into every individual intimacy.

"If history is a precedent, we will enthusiastically embrace these technologies, even as we agonise over whether or not we should do so." (Paul Wolpe)

These breaches of the privacy of individuals, if not duly regulated, might possibly lead to enormous insurance problems, which is another way of stating that it might undermine solidarity between citizens. It might also lead to a loss of social responsibility, if, for example, companies refuse to hire workers with a 'high-risk profile' for mental disorders.

Various sorts of screenings could be proposed to the population, whether on a *public* or on a *private* basis and sometimes on very mercantile grounds. This is very worrying not only for the obvious privacy reasons mentioned above, but also for the impact this could have on the individual's self-identity and on other important aspects of life such as marriage and childbearing. These questions of privacy and of the 'right of not knowing' have already been submitted to discussion with the citizens of many countries in relation to genetic testing, but the actual merging of techniques is magnifying the need for a careful examination of these issues.

Brain-testing juggernaut

The testing of children raises delicate ethical questions, particularly in the current context of increasing social pressures towards 'perfect' and well-performing children. It is obvious that refined diagnostic and predictive techniques can yield valuable aid for paediatricians and psychologists in their tasks, notably with all sorts of slight disabilities and learning disorders. For instance, a child whose brain is not maturing normally could show an unusual pattern while performing some tests. This could make early intervention possible, say with special lessons. But there is a risk of being over-zealous, particularly in view of the current trend towards medicalisation. Should mass screenings of school children be set up? Or should this be left to the decision of the parents? Would they be discredited if they did not take the opportunity to optimise their children's performance? Who would pay for the subsequent increase in the need for specific therapies?

Some observers, like bioethicist Arthur Caplan of the University of Pennsylvania, predict that this could even lead much further: *"Parents might say: 'I want to find out if little Johnny is good at the violin. I'm not going to waste lessons on him if he's got no natural aptitude'. Just as there is an educational-testing juggernaut, there could easily be a brain-testing juggernaut tomorrow."* And he bets on coffee-table books showing the brain patterns of future Mozarts and Einsteins, where parents could try to match up little Johnny's to see where it fits....[35]

One particular point should be made at this stage, i.e. brain-imaging techniques do not deliver objective knowledge. Images and data have to be interpreted by skilled specialists, whose opinions and interpretations frequently vary. No 'hard science', thus, but a skill. As neuroradiologist Mark van Buchem stated: *"Interpreting images is not only recognising a pattern. You also have to understand the pathology. A good radiologist knows what he is looking for."* What would become of the quality of results if critical screenings were left to uncertified persons or to computerised analysis? Moreover, all tests (brain-imaging as well as genetic or biological ones) can sometimes produce incorrect results inherent in the technique, either false positive (finding a disease where none exists) or false negative (overlooking the real case of a disease). This, of course, is not specific to the brain sciences, but it is especially preoccupying in this field because of the psychological weight of an error.

All this new data about our current and future health goes together with the previously debated medicalisation of daily life; it could also increase the general feeling of uncertainty and insecurity since everyone could be exposed to becoming a (future) patient without ever experiencing any symptoms. Whether they want it or not, citizens will have to face the unfamiliar task of living with 'personal' statistical profiles, their burden of anxiety and their potential to influence individual and community-based ways of life. In short: the practice, which in many ways is beneficial, of early and predictive diagnosis might lead to a more individualistic and more anxious lifestyle. This shows the need to educate the public about

---

35 *Ethics and Mapping the Brain*, by Lou Marano, United Press International, 2003.

the relativity of the notion of risk and about the need to clearly define the right of 'not knowing'.

Deviant behaviour

Personal freedom is also at stake when it relates to all kind of deviant behaviour. First of all, it questions the deterministic fatalism that is gaining popularity with the advances in genetics, cognitive neurosciences and brain-imaging. *"As long as our society seeks simple explanations for phenomena as complex as the differences between individuals and groups, danger looms"* (Eric Parens).[36] Deterministic models of human behaviour will always be controversial in Western societies because they challenge the foundational concepts of democracy, namely equality, individual freedom and free will. Moreover, they pose a definition of normality that stigmatises a fringe of the population according to questionable criteria.

But let us assume that various kinds of abnormal, antisocial, sexually deviant or oppositional behaviour could validly be diagnosed or predicted in individuals, what could be the consequences thereof in the realms of privacy and personal freedom? In a law-and-order context or in a very normative society, it could be difficult for individuals to resist the pressure to submit to testing and to escape from the subsequent stigmatisation. It might also become legal to force treatment on criminals, assuming - again - that treatment is available and safe (which is not too unrealistic). The already existing propensity towards court-ordered mandatory treatment would further increase and might no longer be applicable only to guilty 'curable' criminals. All individuals with 'deviant' antisocial behaviour might become liable to some kind of 'revalidation'.

Within the various possible pharmacological and non-pharmacological tools that can be used to act on an individual's behaviour, one has to mention here deep brain stimulation techniques (implanted electrodes) that make it possible to reinforce or suppress certain brain functions, with the advantage of being non-destructive and reversible. This technique has been applied so far for the relief of chronic pain, epilepsy, Parkinson's or Gilles de la Tourette's symptoms, and is being tested for depression and obsessive compulsive disorders, with results described as positive (giving rise to fierce defiance from some patients' associations). But some observers, like the American sociologist Robert Blank, fear that such a technology might also be used for more doubtful reasons. He insists that, once electrodes are implanted, they might allow remote monitoring of brain activity, and he mentions a proposal to use such devices on parolees to monitor their activities through their brain waves. This proposal, he says, raised intense opposition and was immediately abandoned, but this proves *"that even a technique that has credible uses when medically indicated has tremendous abuse potential."*[37]

---

36 'Taking Behavioral Genetics seriously', *Hastings Center Report*, by Eric Parens, 1996.

37 *Brain Policy, How the New Neuroscience Will Change Our Lives and Our Politics*; Robert H. Blank, Georgetown University Press, 1999.

The problem of social control is not a new one, but advances in brain sciences are bringing it to the fore. We are facing the development of an impressive array of techniques which can control or modify behaviour, which offer tremendous opportunities for abuse and which pose serious threats to individual liberty. Even if these innovations can be beneficial to many individuals and sometimes represent their only hope of leading a near-normal existence, society has to establish reasonable guidelines for their use.

Security and judiciary concerns

New developments in brain-reading techniques are also increasingly used for security reasons, and numerous observers point out that it would be wise to set limits to that kind of use in these days of growing paranoia.

An application of brain electro-encephalography is already patented and sold under the name Brain Fingerprinting™. Its principle is to detect the presence of specific pieces of information in the brain of individuals by measuring brain-wave responses to words, phrases, sounds or pictures presented by a computer. If a suspect recognises the details of a scene he has been implicated in, the patterns of his brain waves show that the record of the scene is stored in his brain. This new test is said to be superior to the classical lie-detector and has already been used in some US courts to convict, despite limited testing of the technology for accuracy and reliability. The company that is developing this neurotechnology device is pushing it on the market as a law enforcement tool, a pre-employment screening device (!) and, of course, an anti terrorist screening device. Will we soon see brain-reading devices for airport security?

Another way of establishing the 'truth' could be developed on the basis of the fMRI technique, which is capable of detecting variations in metabolic activity in brain regions important to paying attention and monitoring errors. Such modifications occur with significant accuracy when a subject is lying.

Many scientists, though, warn against extrapolations about one individual's supposed trend or personality from brain images initially meant for research purposes on statistical series. They fear that the high-tech aura and the impact of visual images may lead judges or inquirers to put more weight on evidence from functional imaging than is warranted. They plead for a better public understanding of the capabilities of imaging in order to prevent over-reliance on this source of information.

Freedom of thought

It is nevertheless obvious that these newly developed techniques have the potential to breach the very privacy of a person's own thoughts. Wrye Sententia, Co-Director of the Centre for Cognitive Liberty and Ethics (California, USA) stated in the Amsterdam workshop that this was a threat to personal freedom: *"Without freedom of thought, you have no freedom of speech, and the emerging technologies are changing the parameters of what and how you can think, just as radically as the printing press changed the parameters of what and how you could write, express, or say things. They are to freedom of thought what the printing press was to the freedom of speech."*

Emphasising the idea that freedom of thought is situated at the centre of what it means to be an individual and that it is a core democratic value, she proposed an elaboration that would recognise the need to define explicit protection for brain privacy, autonomy, and choice. Whether or not people find new technologies desirable or acceptable, this would at least secure the rights of 'normal' people to decide reasonably for themselves, when their own psychological/medical mental competency is not at issue.

Legal guidelines on human rights, as with the UN Universal Declaration of Human Rights (Article 18), already emphasise the importance of each person's right to 'freedom of thought', but, according to the analysis of Wrye Sententia, the contours of this concept are not clearly defined and are mainly meant for ensuring freedom of religious opinion. *"This is not enough for an age in which advances in the neurosciences make it possible to monitor and even change thinking."*

Commercial Interests

As a matter of fact, freedom of thought is not only threatened by technical-investigation devices. New techniques derived from breakthroughs in cognitive psychology can be used as well to manipulate our minds, opinions and behaviour softly and imperceptibly.

Proudly claiming its neuroscientific origin is, for example, the very fashionable 'neuromarketing' which integrates brain imaging and neuropsychological testing to assess advertisements or electoral spots. Its promoters claim to offer their clients "unprecedented understanding of how the brain mediates consumer preferences and purchase behaviour, (...) at a conscious and unconscious level"[38]. They are gaining a high profile with feature-length articles published in *Forbes, The New York Times,* and *The Financial Times.* Although many independent experts have expressed doubts about the reliability of these techniques, "this is unlikely to concern marketeers wanting to dazzle potential clients with snazzy imaging technology"[39].

One doesn't know exactly how far these techniques, combined with other marketing 'savoir-faire', will be able to manipulate the purchase behaviour of consumers. *Lancet Neurology* tells about an experiment conducted in 2003 on cola drinkers in order to understand why Coca-Cola continues to outsell Pepsi even though consumers prefer the taste of Pepsi

38 Brighthouse Neurostrategies™ Group, www.brighthouse.com/neurostrategies.html.
39 'The leading edge', in *The Lancet Neurology*, Vol. 3 February 2004.

in blind tastings: it seems that the Coca-Cola brand is so attractive that it overrides what our taste buds are telling us![40]

Another worrying example of the commercial exploitation of new technologies is the device called HyperSonic Sound™, which is able to deliver marketing messages or other noises by using the eardrum to reproduce sounds that cannot be heard around one. In other words, you can hear the voices 'inside your head', like Tom Cruise in the Spielberg movie 'Minority Report', when he walks in a shopping mall and is personally addressed by voices whispering enticing offers inside his ears. Reality has now almost caught up fiction, since Coca Cola (again) is said to be installing in the streets of Tokyo HyperSonic Sound™ devices that will send to the thirsty walkers the sound of ice cubes dropping into a glass and of soda making the delicious "psst" can-opening noise... *"Do we have a right to auditory integrity? Is it ok for us to hear voices in our head that are not our own?"* asked Wrye Sententia, whose institution is leading action in the US for strict regulation of that sort of invasive marketing techniques. *"We already have an onslaught of visual marketing, but you can close your eyes, whereas you can't close your ears!"*. Moreover, creating artificial voices inside the head of mentally fragile persons prone to auditory delusions could probably be responsible for accentuating their disorder.

Military interests
Close to commercial interests, and probably even preceding them, are military interests. A review of the US military agency's latest budget[41] reveals a host of projects aimed squarely at making soldiers smarter, tougher, faster, and stronger - in short, superhuman. For example, they admit that they are currently investigating ways to prevent fatigue and enable soldiers to stay awake, alert, and be effective for up to seven consecutive days without suffering any deleterious mental or physical effects. Remote control of sophisticated machines by thought only, and memory-blunting drugs also come, very logically, within the scope of their research. Much interest is also devoted to the above-mentioned HyperSonic Sound™ devices, that could make very efficient non-lethal weapons delivering 'sound bullets' of 145 decibels, largely above the human threshold of pain.

Influence of the media
Among the questionable manipulations exerted upon our brains, the neurobiologist Jochen Oehler, also present at the Amsterdam workshop, expressed serious concerns about the influence of the media on the development of the brain. He was referring to the growing importance attributable to brain plasticity. Brain plasticity is a relatively new concept which assumes that the brain is a continuously plastic organ: we basically learn all the time, and our brain is constantly being modified by all our experiences. This new theory opens up new concerns, based on the

40 'The leading edge', in *The Lancet Neurology*, Vol. 3 February 2004.
41 'Defense research agency seeks to create supersoldiers', by Bruce Falconer, *The National Journal*, November 10, 2003.

idea that words, images or events can act as sharply as scalpels or as potently as drugs, in the sense that they can durably change our neuronal circuits. The mass media, for example, could thus play a crucial role through day-to-day exposition, and there is a great potential for (mis-)use of its power. Oehler quoted the German philosopher Thomas Metzinger, saying: *"I believe that the media environment created by ourselves is much more dangerous than created pharmacological manipulations."*

Erasing memories

A recent American movie, Eternal Sunshine of the Spotless Mind, is based on a scenario where the main character gets his memories of an unsuccessful love affair erased by some futuristic technology. In real life, some (very common) beta-blocking drugs have shown 'blunting memory' properties as side effects, apparently by reducing the emotional impact of traumatic events. This has raised considerable interest and launched explorative research. The drug called Propranolol is currently being tested in the prevention of post-traumatic stress disorder in accident victims and emergency wards attendees. But there are indeed also many non-medical situations where a memory could wishfully be retained or erased, and not only in unhappy love affairs. If a person is the sole witness to a crime, should the judge be able to require him not to take a forgetting drug until after he testifies? Conversely, could somebody be considered guilty of a crime and be punished for it, if he has erased the memory of it? Could a soldier be held responsible for war crimes if he has received memory-erasing drugs while committing them? As Wrye Sententia pointed it out in Amsterdam, *"this drug raises questions of freedom of thought in the sense that how this will be used and who will be able to use it or not, should lie with the individual and never be compulsory".*

Freedom of information

Lastly, some participants in the Amsterdam meeting raised a real but not specific aspect of personal freedom, which is the right of access to information about all available new technologies. The Internet was particularly at stake, with the incredible hotchpotch of information that it can deliver, from the most rigorous to the most fanciful.

Some participants were in favour of strict control of all information, whereas others denounced the paternalism of such an attitude that infers that the patient is not able to decide with full autonomy what is good for him or her. Jes Olesen wondered *"if we have enough respect for the ability of patients to understand and make decisions on their own behalf"* whereas Ian Ragan, a representative of the pharmaceutical industry, protested against filters being set for information from drug companies for patients: *"If you are going to have informed patients and they are going to have access to plenty of the snake-oil stuff that can be found on the Internet, I think they should have equal access to the information from the pharmaceutical industry and be allowed to make up their own minds."* More of this is developed in the section 'Dialogue'.

Ethics of research

Conducting active research upon the human brain is a rather recent affair. Until functional brain imaging opened visions of the living brain, neurobiologists had to wait with crossed fingers until some unfortunate patient walked through their door with an injury or disease that affected the region or function which they were trying to study. Clinicians, on their side, were doing their best to alleviate suffering on empirical grounds and with symptomatic means. Now this is radically changing. Brain activity can be observed, measured, and modified. This opens up new and unhoped-for potentialities, but also great danger as regards ethical side-slipping.

For the most part, the ethics of the clinical brain sciences can be superimposed on traditional bioethics. It includes familiar issues like optimal clinical trial design, guidelines for the use of fetal tissues or stem cells, privacy rights to results and so on. However, a very particular aspect of clinical research in brain science is that informed consent must be expressed by the damaged organ itself! This question of 'sound mind consent' is particularly delicate when dealing with people at the fringes of consciousness, in every sense of the word: comatose persons, persons suffering neurodegenerative diseases, or severe psychiatric patients. It is also ordinarily taken for granted that persons are able to choose what is in their best interest. But the very concept of informed consent can be twisted in conditions where persons are not able to take such decisions, as in deep depression or where they are driven by some desperate hope to accept clinical trials with dubious benefits for themselves as, for example, in the first stages of a degenerative disease.

Safety and the precautionary principle
The first psychotropic drugs were discovered in the fifties. This means that we don't have a very long hindsight as regards assessing their long-term effects. This is, of course, also the case with many other pharmaceutical classes, but the particularity of psychotropic drugs is that they may, not exceptionally, be taken lifelong.

Thus, it might be possible that unexpected effects pop up with long-term use of the new drugs, knowing that some of these substances are potentially able to cause permanent changes in the synaptic connections (just as illegal drugs do, but also as memory does by continuously remodelling the neuronal interconnections). The particular case of children is worth pointing out, since the trend is upwards to medicalise them from early childhood on.

On the other hand, we don't know either how an untreated mental disease might affect the brain in the long term. Nor an individual's life trajectory. So finally, what is worse for diseased people: the long-term effects of psychotropic drugs or not employing them at all? Can we quantify the disadvantages of (as yet unidentified) side effects against the price of not treating patients?

More worrying is the fact that the more the new drugs become devoid of immediate side effects, the more they will escape the medical control of prescription. All fanciful smart drugs might thus be freely available and heavily advertised greatly in advance of careful and complete research. Commercial pressure but also patient demand could accentuate this phenomenon. Nobody knows what this can lead to. As Demosthenes Agrafiotis humorously remarked, *"we don't know the long-term impact of hyperknowledge!"*

Ethics of information

The preceding issues are very classical ones and even if they are especially acute in the context of the brain sciences, they are not consistently different from the usual clinical ethics. As Bernard Reber said in the Amsterdam workshop: *"Many ethical questions surrounding brain science are not new. But there are 'metaethics': some assumptions of brain research lead to a more deterministic general world view, and that shift might have negative consequences for public appreciation of individual freedoms and choices"* (see the section on 'reductionism' for an elaboration of this). For this reason, Bernard Reber insisted that precautions in current research presentations should be considered tremendously important, because partial understanding of complex phenomena could lead the public - and the policymakers - to completely wrong opinions. *"We have a kind of epistemic and semantic duty before speaking of any consequences in a field which is so heterogeneous and so fantasmatic as that of the brain sciences..."*

"We should acknowledge that science is not only about making facts, but also about making cosmologies, that is stories about who we are. This creates a possibility for public debate." (Andreas Roepstorff)

The importance of communication about science is such that it will be further developed in a separate last section, but the ethical aspect of this issue has to be stressed here. It is first of all a question of fairness: just as clinicians have to make sure that their words are correctly understood by their patients in order to prevent the risk of false hopes, inappropriate despair or self-fulfilling prophecies, researchers (from the public as well as from the private sectors) should insist that communication be made completely and honestly about the results of their work, in order to prevent the occurrence of oversimplified views or unrealistic expectations among the public. Ian Ragan of the multinational pharmaceutical company Eli Lilly: *"Unrealistic expectations have been exacerbated by governments demanding a short-term return on investments in order to justify spending taxpayers' money - and this is why it is a societal issue as regards research. There has been exaggeration by scientists who need to get grants to do their work and of course there has been misinformation in the media."*

But the real 'metaethics' are somewhere else. Andreas Roepstorff: *"Science - and at the moment particularly brain science - is not only about making facts. It is also about making cosmologies, it is about making stories about who we are, what our place in the world is. And so to play out discussions that relate not only to science but to much wider discussions."*

Interdisciplinarity
Therefore, it is very important that developments in the brain sciences be firmly embedded in society. Up till now, the ethical aspects of science have generally been tackled by adding post factum an 'ethics' component to the research projects. But this has not really led to major implications for the way knowledge has been acquired; rather it became a sort of philosophical damage control that came after the facts. Roepstorff pleaded for the idea that the brain should not be studied through the narrow views of hyper-specialised disciplines, but through the wide angle of interdisciplinarity: *"One way you can make facts robust is by making certain that interdisciplinarity is already there at the level of research. Interdisciplinarity means not only different types of brain scientists, but also that you have philosophers, anthropologists and psychologists sitting on the bench. Because that opens up important discussions at a point before the facts are actually being made. That insures that the community of trust that the public has to believe in becomes larger."*

If interdisciplinarity is established from the funding step of a research project, it will make certain that the embedding of science in society takes place early on. Which Berit Faber further emphasised by proposing the formula 'just-in-time ethics', to name the need for public debates being started not too early and not too late.

Animal rights
As in many other scientific disciplines, animals are used in the process of gathering knowledge concerning the brain. With the use of animals for human needs, there are always questions of proportion ('is the suffering of animals balanced out by the expected gain for humans?') and inevitability ('is it necessary to use animals, or are there other ways to find out what we want to know?'). Opinions differ concerning the answers to these questions.

One field in which animals are extensively used is in psychopharmacology. Psychotropes are usually tested on animals before they are tested on humans, in order to know if the molecule works and to know if serious side effects will occur. Many people feel that the above questions of proportion and inevitability can be answered positively in the case of the development of new drugs for brain diseases. If these same drugs are being used for reasons of enhancement, the answers to these questions might be less straightforward.

A small but determined group is against any use of animals for human needs whatsoever. According to Ian Ragan, *"the ability to do research is threatened by animal rights extremists"*, because multinational pharmaceutical companies might prefer to place their research units outside Europe, where the chances of militant actions against animal-testing are lower. This might come to a serious loss for the European research community.

Animals are also used in more fundamental brain research, mainly in the cognitive neurosciences. For example, the important fact that mammals exposed to stimulating environments from birth on have far more connections between brain cells than genetically identical mammals raised in seclusion, was first established in the case of rats. The implication of this research is enormous. It shows that most of the brain gets built after birth and that the brain uses experience from the outside environment to form its circuits of thinking, memory, emotions, etc. This insight might lead to more public concern for the quality of the environment of small children. But work in this direction could only be started by killing rats and actually counting the number of connections.

The inevitability of the use of 'furry test tubes'[42] in order to find out about the brain is beyond dispute, but the question of proportion is more difficult to answer when it comes to fundamental research, since the expected gain is less clear. And there is another catch. In order to find out more about 'higher', cognitive processes in the brain (and this is where a lot of research in cognitive neurosciences is heading towards), we have to use 'higher' animals - precisely for the reason they are similar to us. In the words of Donald Kennedy, editor-in-chief of Science: *"It may not change our view of how we deal with animals as experimentalists, but it will certainly change our view of the continuity of all living things"*.[43] Many people feel that it is ethically different to use a fruit fly during research, or a cat or an ape. During the Amsterdam workshop, cognitive neuroscientist Richard Morris commented: *"If we are to get rid of the research on non-human primates for ethical reasons, we will be left with a huge gap between the work which is done on rodents and the application of this research to humans. We would have to live with that gap and with all of its implications."*

The German philosopher Thomas Metzinger complicates this ethical question a bit more.[44] For him, the result of brain science itself makes it almost inevitable to concede that higher animals have feelings as well as a (non-cognitive) sense of 'knowing' they are their body. In other words, *"they have the central capability to experience their suffering as their own suffering"*. This makes it even harder to balance the interests of higher animals against the very exciting and promising tracks brain scientists are exploring in using their 'furry test tubes' - though we should not forget that the same problem is at issue in our meat industries.

---

42 A phrase Richard Morris used during the workshop.

43 *Neuroethics: Mapping the Field*, p. 201.

44 In his article 'Der Begriff einer Bewusstseinskultur'. In G. Kaiser (Hrsg.), *Jahrbuch 2002/2003 des Wissenschaftszentrums Nordrhein-Westfalen. Düsseldorf.*

Dialogue and information exchange

In a media world more and more ruled by sound bites and scoops, it is hard for the lay public to put opportunities and threats raised by developments in the brain sciences in proper perspective. Headlines in the newspapers and items on television shows might give rise to-as yet - unreasonable hopes of a fast and complete cure for neurological and psychiatric diseases. Also, some media present a scary picture of a future Brave New World, peopled by placid, robot-like citizens and consumers as the inevitable outcome of possibilities opened up by the brain sciences.

One of the aims of the Amsterdam workshop was to prepare citizen panels that will exchange ideas on the impact of the brain sciences on our culture. Of course, the methodology of a European discussion among lay people concerning such complex issues is very challenging. The Amsterdam workshop was meant to deliver content for these discussions, not to think through the ways in which these discussions should be organised, as other specialists are already attending to that task. That being said, the Amsterdam workshop made it quite clear that the position of scientists within the community and the way information exchange between scientists and public is managed is a hot topic for scientists as well as stakeholders.

Science and public media
Every participant underlined the importance of a scientifically well-informed public, given the speed of scientific progress and *"the need, in a democratic society, to involve the public in the decisions of politicians and commerce about how science should be applied"*, as Colin Blakemore, head of the UK Medical Research Council, puts it.[45]

Still, it is difficult for scientists to get their ideas and findings well represented in the media. There are several reasons for this. One is the logic of the 'science market'. Academic scientists are sometimes seduced into exaggerating their findings in order to get grants to continue their work. Industry research is communicating assertively in order to boost the share values of the companies. Since 'a major breakthrough' is what media most love to cover, this leads to sweeping claims which later have to be nuanced or corrected - thereby undermining the trust of the public in the 'facts' of science. Ian Ragan: *"A fundamental discovery normally does not even lead to treatment within half a lifetime. But we don't like to say that. Not as a researcher, because it might block your grant, and not as a government, because you want to explain to your voters why you are spending money on biomedical research."*

The fact that news about science increasingly comes from someone in a PR-role, as opposed to someone in a research role, also makes it difficult for the public to distinguish between proper science education and the

45 *Neuroethics: Mapping the Field*, p.213-223.

marketing of science. As Ron Kotulak, a biologist and science writer, states: *"scientists who are reporting new findings should be available to the media for interviews or background information"*.[46] It might also help if universities changed their systems for the accreditation of scientists, so that communication with the public is formally credited.

Secondly, the 'character' of the scientist is ill-at-ease with the logic of the media. A scientist among his or her peers strives to appear balanced and rational, highlights procedures and facts, and does not want to draw much attention to any personal involvement in the issue at stake. This attitude does not 'sell' well to the (mass) media which have a tendency to focus on motives and emotions. Good science is, by nature, a conservative enterprise that moves along very carefully. It is hard work to make good 'infotainment' out of such a process, and not many scientists (or media people) have the required background or inclination to do so.

Thirdly, mainstream media do not really offer a good podium for science. Often journalists are conditioned to 'present both sides' of an issue, thereby inviting a scientist to debate with, for example, an astrologist or an activist. According to Colin Blakemore, *"single-issue groups have exploited the concern of the press to be balanced"*.[47] Of course scientists should be available for public debates on controversial scientific issues - as long as both sides are legitimate sides. This is not always the case. Scientists debating with persons proposing 'snake-oil' remedies often find, to their frustration, that the status of the 'knowledge' of both parties is presented as equal.

Lastly, the tendency of the media to focus on 'newsworthy' or even 'sensational' facts that science may produce blurs public appreciation of what science actually is. Scientists have a task here too: they should not so much report more scientific facts as try to educate the public more about what science is. That is: they should present science as a process. Scientists might, for example, communicate their uncertainties in order to show that science emerges from the turmoil of conflict and controversy. As Jan Willem Berkelbach van der Sprenkel said at the Amsterdam workshop: *"We should convey to the public that we are using simplifications to answer our questions, that it is not the complete answer."* Scientists might also convey more that brain research on controversial topics such as paedophilia does not necessarily mean a justification of these topics. Paolo Valerio put this as follows: *"We should help the public to understand that explaining and guiding do not mean justifying."*

An educated public is more likely to be supportive of science-based policies, instead of non-proven 'alternative' solutions. And as the public learns that scientific facts do not have an irrefutable, uncontested status, they might feel less deceived if the scientific 'fact' of yesterday has to be corrected today. A scientifically informed public might also have a better capacity

46 *Neuroethics: Mapping the Field*, p.233.
47 *Neuroethics: Mapping the field*, p.259.

to assess risks in their own lives, because they will accept that scientific knowledge comes in terms of probabilities rather than absolute certainties. At the moment this is hard, Ian Ragan finds, since *"society is increasingly risk-averse and litigious"*. To his mind, this demand for absolute certainty and safety is stopping effective research.
The advantages of another type of public debate on scientific issues are fairly obvious. It is much less obvious how this should come about, given the logic of the scientific world and the practices of the media. But as far as scientists are concerned, it will surely mean paying attention to the public and making an effort.

Scientists versus patients
As regards dealing with the public in their role as *patients*, scientists could surely do better, according to the representatives of the patients present at the Amsterdam workshop. They ventilated their feeling that scientists and doctors often consider communication as a one-way street. According to Paolo Lucio Morselli, many mentally ill patients *"work on misinformation and wrong hopes"*. Patients are seldom asked for their views, whereas a properly informed patient can make reasonable and realistic plans for him- or herself. This is the experience of Paolo Lucio Morselli: *"A well-informed patient helps to reduce care costs and so could help in saving a lot of money."*

Alistair Newton also states that patients need to be involved in the management of their condition. This might mean that different interests of patients and scientists come to the fore. Scientists typically want to cure. Take for example Parkinson's disease: Parkinson's patients themselves might be *"more interested in their bowel function and quality-of-life issues, which are not addressed very much by science"*, according to Alistair Newton. *"Basic scientists should see that they need to consider carefully who and what it is they are working for"*, he warns.

Brain sciences: an interdisciplinary field...
The above-mentioned statements about the relationship between scientists and the lay public and scientists and patients are relevant to, but not typical of, the brain sciences. There are two aspects of communication and information exchange that are typical for the brain sciences. One is the extraordinary interdisciplinary character of the field of the brain sciences. The brain sciences are at the crossroads of many traditional disciplines: medicine, biology, psychology, philosophy. This places demands on the scientists from different disciplines participating in debates on the brain sciences, as the Amsterdam workshop itself demonstrated. Scientists working in this field come into contact with many different, and often unfamiliar, scientific vocabularies. In this situation, it is not easy to get one's facts right. As the philosopher Bernard Reber stated: *"If you don't want to do science-fiction ethics, you need to know precisely where the problems are"*, and this is particularly hard in a new and interdisciplinary field like the brain sciences. Still it is a task for scien-

tists to see where the field is heading. Demosthenes Agrafiotis pointed out that brain scientists need to concern themselves with the 'dynamics' of the field in order to operate in a responsible way. Only then will it become possible *"to organise these dynamics in order to prevent one group from keeping this territory as its own"*.

...with potentially unsettling results

The second aspect in respect of which communication concerning the brain sciences is especially delicate, is that scientific findings are potentially very unsettling. The brain sciences affect human bodies and mind directly. This leads to public fear of loss of control and loss of meaning. Brain scientists have to deal with this. As Jan Willem Berkelbach van der Sprenkel puts it: *"We are not only giving a statistically significant answer to the public, we also have to acknowledge the impact of the answer that we give"*.

So, by studying the human brain one enters a particularly emotionally-charged field. Michael Gazzaniga, who as a cognitive neuroscientist has a seat on the President's Council on Bioethics, reports: *"You have to learn all these different vocabularies. You're a knowledgeable scientist, but it turns out you don't know much. You have to learn that it is normal for people to have strong beliefs"*.[48]

As stated previously in this report, recent findings in the brain sciences might well make it necessary in the eyes of scientists to reconsider seriously the status of concepts like free will, responsibility or the soul. But scientists cannot expect the public to see this need immediately or, if they do, to happily attend to this need. These are typically concepts about which many people will hold 'strong beliefs' which they might find very unsettling to discuss. Scientists will have to acknowledge this in their dealings with the public. As Flavio Keller pointed out: *"As scientists we cannot get out and say: 'There is no soul'. The soul has philosophical, religious and other implications for many people and you cannot just say that these lack any basis. It's not respectful. We are not entitled to this kind of affirmation on the basis of our knowledge."*

This paragraph started with the statement that the Amsterdam workshop was not about designing the discussions with and among European citizens. Nevertheless, a few things can be noted on the basis of this workshop. A discussion of this type can be successful only if the debate becomes a two-way process of interaction between scientists and the public, rather than a one-way didactic presentation of the truth by scientists to ordinary people.

And there is hope that this can be done, *"if"*, as Berit Faber puts it, *"we strive towards transparent procedures for scientists and the industry in the*

---

48 *Neuroethics: Mapping the field*, p.241.

way we decide about things." Former experience in Denmark with citizens' panels make her confident that lay people and young persons who have informed ideas about the scientific process will be able to discuss the social and ethical implications of scientific findings. *"They will be able to bring something into scientific circles"*, she assured the audience.

Appendix 1

| Glossary |

Brain fingerprinting

Brain fingerprinting is designed to determine whether an individual recognizes specific information related to an event or activity by measuring electrical brain wave responses to words, phrases, or pictures presented on a computer screen. The technique can be applied only in situations where investigators have a sufficient amount of specific information about an event or activity that would be known only to the perpetrator and investigator. In this respect, brain fingerprinting is considered a type of Guilty Knowledge Test, where the 'guilty' party is expected to react strongly to the relevant details of the event or activity. Existing (polygraph) procedures for assessing the validity of a suspect's 'guilty' knowledge rely on measurement of autonomic arousal (e.g., palm sweating and heart rate), while brain fingerprinting measures electrical brain activity via a fitted headband containing special sensors. Brain fingerprinting is said to be more accurate in detecting 'guilty' knowledge distinct from the false positives of traditional polygraph methods, but this is hotly disputed by specialized researchers.

Adapted from The Center of Cognitive Liberty and Ethics,
http://www.cognitiveliberty.org/

Brain plasticity

Brain plasticity, or neuroplasticity, is the lifelong ability of the brain to reorganize neural pathways on the basis of new experiences. As we learn, we acquire new knowledge and skills through instruction or experience. In order to learn or memorize a fact or skill, there must be persistent functional changes in the brain that represent the new knowledge. But the natural ability of the brain to reorganize also plays a major role in compensating for injury, damage or degeneration.

Brain sciences, a working definition

Brain sciences is the field of study which deals with the function of the brain, in all aspects, notably neurological, psychical and cognitive aspects. The brain can be approached through structural, physiological, psychological, biochemical, genetic, developmental, or pharmacological ways, without leaving aside the newly rediscovered role of subjectivity, emotions and affects The study of the brain can operate at different levels, from the molecular interactions of receptors and proteins, to the study of behaviour of individuals, and even of populations.
It is a global field of which interdisciplinarity is the cornerstone.

Brain sciences, by essence, are central to the definition of the human mind. Therefore they are closely linked to all human sciences, notably philosophy, anthropology, psychology, pedagogy, sociology, theology. New developments are now making bridges between brain sciences and more peripheral fields of knowledge like economy or marketing, and probably others still to come.

Through their future developments, brain sciences bring the potential to deeply modify the society and the vision of what it is to be human.

Computed tomography

Computed axial tomography (CAT), computer-assisted tomography, computed tomography, CT, or body section roentgenography are different names for the process of using digital processing to generate a three-dimensional image of the internals of a patient from a series of two-dimensional x-ray axial images. The individual x-ray axial slice images are taken using an x-ray tube that rotates around the object taking many scans as the object is gradually passed through the gantry. The multiple scans from each 360-degree sweep are then processed to produce a single cross-section. The word 'tomography' is derived from the Greek *tomos* (slice) and *graphia* (describing). It is used in medicine as a diagnostic tool and as a guide for interventional procedures. Sometimes contrast materials such as intravenous iodinated contrast are used. This is useful to highlight structures such as vessels or intestines that otherwise would be difficult to delineate from their surroundings. Using contrast material can also help to obtain functional information about tissues.

Adapted from Wikipedia, http://en.wikipedia.org/wiki/Ct_scan

Consilience

The term consilience, or the Unity of knowledge (literally a 'lumping together' of knowledge), was first mentioned in *The Philosophy of the Inductive Sciences*, by William Whewell in 1840. In this synthesis Whewell explained that, "The Consilience of Inductions takes place when an Induction, obtained from one class of facts, coincides with an Induction, obtained from another different class. Thus Consilience is a test of the truth of the Theory in which it occurs." The Scientific method has become almost universally accepted as the exclusive method for testing the status of any scientific hypothesis or theory. 'Inductions' which arise out of applications of the scientific method are, by definition, the only accepted indicators of consilience. *Consilience: The Unity of Knowledge* is also the title of a 1998 book by Edward Osborne Wilson. In this book, Wilson discusses methods that have been used to unite the sciences and might in the future unite the sciences with the humanities. Wilson prefers and uses the term consilience to describe the synthesis of knowledge from different specialized fields of human endeavour. In the last chapter of his book Wilson describes the relationship between genes and culture. The basic element of culture is the meme. When a meme exists in a brain, it has the form of a neuronal network that allows the meme to function within semantic memory. The link between genes and culture is that our genes shape our brains (in cooperation with the environment) and our brains allow us to work with memes as the basic units of culture.

Adapted from Wikipedia, http://en.wikipedia.org/wiki/Consilience and http://en.wikipedia.org/wiki/Consilience:_The_Unity_of_Knowledge

Deep brain stimulation

Deep brain stimulation (DBS) is a surgical procedure originally used to treat severe essential tremor and tremor, rigidity and bradykinesia (slow movement) associated with Parkinson's disease. The surgery involves the implantation of an electrode to the thalamus of the brain powered by a battery-operated neurotransmitter placed under the collar bone. The electrode targets the substantia nigra in interfering with the electrical signals that cause tremors, rigidity or bradykinesia. After the surgery is completed, an expert calibrates the unit in order to maximize its effectiveness. Typically DBS is performed unilaterally on the side of the brain most debilitated by the disease, but in some cases it is performed bilaterally in a single operation.

DBS has been found to significantly alleviate symptoms in two thirds of Parkinson's patients. Recently deep brain stimulation has been used to treat a few patients with Gilles de la Tourette syndrome and patients with severe and treatment resistant obsessive compulsive disorder.

Adapted from Wikipedia,
http://en.wikipedia.org/wiki/Deep_brain_stimulation

Electro- and magneto-encephalography

Electroencephalography is the neurophysiologic exploration of the electrical activity of the brain by the application of electrodes to the scalp. The resulting traces are known as an electroencephalogram (EEG) and represent so-called brain waves. This device is used to assess brain damage, epilepsy and other problems. Historically four major types of brainwave are recognized (alpha, beta, delta and theta). Beta is the frequency range above 12 Hz. It is often associated with active, busy or anxious thinking and active concentration.

Alpha is the frequency range from 8 Hz to 12 Hz. It is often associated with a relaxed awake mind, daydreaming, watching TV or the first stage in falling asleep. Theta is the frequency range from 4 Hz to 8 Hz and is associated with REM sleep, creativity, visual imagery, reverie and recollection. Delta is the frequency range below 4 Hz and is often associated with deep sleep.

Magneto-encephalography (MEG) is the measurement of the magnetic activity of the brain, usually conducted externally, using extremely sensitive devices such as SQUIDs. Because the magnetic signals emitted by the brain are very weak, shielding from external magnetic signals, including the Earth's magnetic field, is necessary. An appropriate magnetically shielded room is very expensive, and constitutes the bulk of the expense of an MEG system.

MEG is a relatively new technique that promises good spatial resolution and extremely high temporal resolution, thus complementing other brain activity measurement techniques such as Electroencephalography (EEG), Positron emission tomography (PET), and functional Magnetic Resonance Imaging (fMRI).

Adapted from Wikipedia,
http://en.wikipedia.org/wiki/Electroencephalography and
http://en.wikipedia.org/wiki/Magneto-encephalography

Eye movement desensitisation and reprocessing

Eye movement desensitisation and reprocessing (EMDR) is a relatively new form of psychotherapy developed by Dr Francine Shapiro. The therapy is employed to treat post traumatic stress disorder and is based on behavioural techniques, in which the patient makes rhythmic eye movements while imagining the traumatic event. Until now, no one has been able to adequately explain how EMDR is supposed to work. Some practitioners claim that the eye movements are designed to stimulate the information-processing system in the brain. The aim of the treatment is to help along the processing of the traumatic events, and speed up readjustment and recovery.

Gene therapy

An experimental procedure aimed at replacing, manipulating, or supplementing nonfunctional or misfunctioning genes with healthy genes. Although the technology has been in development for 15 years, it has only been used with minor success so far.

In theory, it is possible to transform either somatic cells (most cells of the body) or cells of the germline (such as germ stem cells, sperm and eggs). All gene therapy so far in people has been directed at somatic cells, whereas germline engineering in humans remains only a highly controversial prospect.

Somatic gene therapy can be broadly split into two categories: *ex vivo* and *in vivo* gene therapy. In *ex vivo* gene therapy, cells are taken out of the body, genetically modified and then transplanted back. In *in vivo* gene therapy, cells are genetically changed while still in the body. Until now, most successful clinical trials with gene therapy have been *ex vivo* procedures. *In vivo* approaches have largely been unsuccessful because it is very difficult to introduce the correcting gene into the right target cell and to get it properly expressed and propagated.

Genetic engineering

Genetic modification, genetic manipulation or genetic engineering are terms for the process of manipulating genes in an organism, usually outside of the organism's normal reproductive process.

It often involves the isolation, manipulation and reintroduction of DNA into model organisms, usually to express a protein. The aim is introduce new genetic characteristics into an organism to increase its usefulness such as introducing a novel characteristic in the model organism, producing a new protein or enzyme, or increasing the yield of a crop species. Examples are the production of human insulin through the use of modified bacteria and the production of new types of mice for research, through genetic redesign.

Adapted from Wikipedia, http://en.wikipedia.org/wiki/Genetic_engineering

Hypersonic sound

In hypersonic sound the music or voice from the audio source is converted to a highly complex ultrasonic signal and emitted into the air in a directional way. The sound forms a virtual column of sound directly in front of the emitter, much like the light from a laser. The hypersonic sound itself cannot be heard by the human ear, but all along the column of ultrasonic sound, the air creates new sounds (the sound that was originally converted to an ultrasonic wave). Since the sound is only created right in the column of ultrasonic energy, it does not spread in all directions like the sound from a conventional loudspeaker; instead it stays locked tightly inside the column of ultrasonic energy. So only someone within the column of hypersonic energy is able to hear the sound.

More information can be obtained from American Technology Corporation, www.atcsd.com

Magnetic resonance imaging

Magnetic resonance imaging (MRI) is a method of creating images of the inside of organs in living organisms. The original name for the technology was nuclear magnetic resonance imaging (NMRI), but the term 'nuclear' was dropped because doctors and scientists thought that it carried negative connotations from its usage in other contexts.

The principle of MRI is based on the differential magnetic properties of molecules in cells and tissues. The technique most frequently relies on the relaxation properties of magnetically-excited hydrogen nuclei in a watery environment (like in the cells of our body). The patient is briefly exposed to a burst of radiofrequency energy, which in the presence of a magnetic field, puts the nuclei in an elevated energy state. As the molecules undergo their normal, microscopic tumbling, they shed this energy back to their surroundings, in a process referred to as 'relaxation'. Differences in relaxation rates are the basis of MRI images—for example, the water molecules in blood are free to tumble more rapidly, and hence, relax at a different rate than water molecules in other tissues. Different scan sequences allow different tissue types and pathologies to be highlighted. A contrast agent is sometimes injected in the sample to augment these differences and improve sensitivity.

These relaxation signals are recorded by sensors surrounding the patient and the resulting data are processed by a computer to generate an image of the tissue. Thus, the examined tissue can be seen with its quite detailed anatomical features. In clinical practice, MRI is used to distinguish pathologic tissue such as a brain tumour from normal tissue.

In MRI various signals can be picked up by the sensors. T1-weighted MRI scans rely on relaxation in the longitudinal plane, and T2 weighted MRI scans rely on relaxation in the transverse plane.

Functional MRI (fMRI) measures signal changes in the brain that are due to changing neural activity. The brain is scanned at low resolution but at a rapid rate (typically once every 2-3 seconds). Increases in neural activity cause changes in the MR signal via a mechanism called the

BOLD (blood oxygen level-dependent) effect. Increased neural activity causes an increased demand for oxygen, and the vascular system actually overcompensates for this, greatly increasing the amount of oxygenated haemoglobin relative to deoxygenated haemoglobin. Since deoxygenated haemoglobin reduces MR signal, the vascular response leads to a signal increase that is related to the neural activity. The precise nature of the relationship between neural activity and the BOLD signal is a subject of current research.
Reflecting the fundamental importance and applicability of MRI in the medical field, Paul Lauterbur and Sir Peter Mansfield were awarded the 2003 Nobel Prize in Medicine for their discoveries concerning MRI. Lauterbur discovered that gradients in the magnetic field could deliver two-dimensional images. Mansfield analyzed the gradients mathematically.

Adapted from Wikipedia, http://en.wikipedia.org/wiki/MRI

Model organism

A model organism is an organism (bacterium, plant, animal) that is extensively studied with the aim of understanding particular biological phenomena, with the expectation that discoveries made in the model organism will provide insight into the workings of other organisms, like humans. This paradigm is based on the notion that evolution reuses fundamental biological principles and conserves metabolic, regulatory, and developmental pathways. So, very often, biological mechanisms studied in mice, or even worms, fruit flies or yeasts, can be extrapolated to humans.
There are many model organisms. Yeasts, particularly *Saccharomyces cerevisiae* ('baker's' or 'budding' yeast), have been widely studied, largely because they are quick and easy to grow. The cell cycle in a simple yeast is very similar to the cell cycle in humans. Both are regulated by the same proteins. The fruit fly *Drosophila melanogaster* was studied, again because it was easy to grow for a multicellular organism. The roundworm *Caenorhabditis elegans* is studied because it has very stereotyped development patterns and can be easily monitored of for abnormalities.
Rodents, especially mice and rats, are also widely studied because many disease mechanisms overlap between rodents and humans. Moreover, scientists have found ways to genetically engineer mice and to add genes to them (transgenic mice) or to make one or more genes in its chromosomes inoperable or 'knocked out'. Mice are the laboratory-animal species most closely related to humans in which the knockout technique can be easily performed. Therefore knockout mice are frequently used in studies on human disease mechanisms. Since identical or nearly identical versions of most human proteins exist also in mice, knocking out the gene that encodes the related enzyme in mice enables researchers to mimic the effect of the pathology at the molecular level in the context of a living animal.

Adapted from Wikipedia, http://en.wikipedia.org/wiki/Model_organism and http://en.wikipedia.org/wiki/Knock-out_mouse

Molecular biology

Molecular biology is the study of the structures and properties of molecules found in the cell. The field overlaps largely with other areas of biology, particularly genetics, biochemistry and cell biology. Molecular biology chiefly concerns itself with understanding the interactions between the various systems of a cell, including the interrelationship of DNA, RNA and proteins and learning how these interactions are regulated. Recently the study of the genome, the assembly of all the genes of an organism, and its function has been amongst the most prominent subfields of molecular biology.

Nanotechnology

Nanoscience and nanotechnology involve studying and working with matter on an ultra-small scale. One nanometre is one-millionth of a millimetre. A single human hair is around 80,000 nanometres in width. Nanoscience and nanotechnology encompass a range of techniques rather than a single discipline, and stretch across the whole spectrum of science, incorporating medicine, physics, engineering and chemistry. Some believe that nanotechnology is the next big thing to emerge from science and engineering and that it could offer us tremendous benefits. Scientists, for example, are investigating whether nanotechnology could be used to improve the delivery of drugs and are investigating whether nanoscale carbon could be used to increase the speed and power of computer circuits. Due to the small size at which nanotechnology operates, physical phenomena not observed at the macroscopic scale dominate. These nanoscale phenomena include quantum effects and short range forces such as van der Waals forces. Furthermore the vastly increased ratio of surface area to volume promotes surface phenomena.

Adapted from Wikipedia, http://en.wikipedia.org/wiki/Nanotechnology and 'Nanoscience and nanotechnology', The Royal Society, http://www.royalsoc.ac.uk/nanotechnology.

Neurotrophic factors

Neurotrophic factors are proteins that regulate the growth, differentiation and survival of neurons in the nervous system. These proteins are also involved in synapse formation. One of the most widely studied neurotrophic factors is brain derived neurotrophic factor (BDNF). It is active predominantly in the hippocampus, cortex, and basal forebrain-areas vital to learning, memory, and higher thinking.

Although the vast majority of neurons in the mammalian brain are formed prenatally, parts of the adult brain retain the ability to grow new neurons from neural stem cells; a process known as neurogenesis. Neurotrophins are chemicals that help to stimulate and control neurogenesis, BDNF being one of the most active.

Adapted from Wikipedia,
http://en.wikipedia.org/wiki/Brain-derived_neurotrophic_factor

Pharmacogenetics

The environment, diet, and general state of health can all influence how a person responds to medicines. But perhaps the key factor is genetics. The study of how individual people respond to medicines due to their genetic inheritance is called pharmacogenetics. The ultimate goal of pharmacogenetics is to understand how someone's genetic make-up determines how well a medicine works in his or her body, as well as what side effects are likely to occur. In the future, advances gleaned from pharmacogenetics research will provide information to guide doctors in getting just enough of the right medicine to a person—the practice of 'personalized medicine'.

Positron emission tomography

Positron emission tomography (PET) is a medical imaging technique in which radioactive 'tracer' isotopes are injected into a living subject (usually in the blood circuit). The isotope emits positrons which 'fuse' with electrons. As a result of this 'fusion', gamma-rays are produced which can be detected by a camera. PET scanning is invasive, in that radioactive material is injected into the patient. However the total dose of radiation is small.

A PET scan allows the visualisation and the relative quantification of the changes of regional blood-flow in various anatomical structures, including the brain. PET as a technique for scientific investigation is limited by the need for clearance by ethics committees to inject radioactive material into subjects, and also by the fact that it is not advisable to subject a person to too many scans. Furthermore, due to the high costs of cyclotrons needed to produce the short-lived radioisotopes for PET scanning, few hospitals and universities are capable of performing PET scans.

PET scanning of the brain is based on an assumption that areas of high radioactivity are associated with brain activity. What is actually measured indirectly is the flow of blood to different parts of the brain, which is generally believed to be correlated, and usually measured using the tracer oxygen-15.

Adapted from Wikipedia, http://en.wikipedia.org/wiki/Positron_Emission_Tomography

Psychotherapy

Psychotherapy is a set of techniques believed to cure or to help solve behavioural and other psychological problems in humans. The common part of these techniques is direct personal contact between therapist and patient, mainly in the form of talking. Psychoanalysis was the original form of psychotherapy, but many other theories and techniques are now used by psychotherapists. A distinction can be made between those psychotherapies that employ a medical model and those that employ a humanistic model. In the medical model the clients are seen as unwell and the therapist employs his/her skill to help them back to health. In the humanistic model the therapist facilitates learning in the individual

and the clients' own natural process draws them to a fuller understanding of themselves.
One particular form of psychotherapy is cognitive behaviour therapy. It is a kind of psychotherapy used to treat depression, anxiety disorders, phobias, and other forms of psychological disorder. It involves recognising distorted thinking and learning to replace it with more realistic substitute ideas. Cognitive therapy is often used in conjunction with mood stabilizing medications.
Cognitive therapy is not an overnight process. Even once a patient has learnt to recognise when and where his thought processes are going awry, it can take months of concerted effort to replace an invalid thought with a more suitable one. But with patience and a good therapist, cognitive therapy is considered to be a valuable tool in recovery.

Adapted from Wikipedia, http://en.wikipedia.org/wiki/Psychotherapy and http://en.wikipedia.org/wiki/Cognitive_therapy

Stem cells

Stem cells have the remarkable potential to develop into many different cell types in the body. Serving as a sort of repair system for the body, they can theoretically divide without limit to replenish other cells as long as the person or animal is still alive. Stem cells have two important characteristics that distinguish them from other types of cells. First, they are unspecialised cells that renew themselves for long periods through cell division. The second is that under certain physiological or experimental conditions, they can be induced to become cells with special functions such as the beating cells of the heart muscle, the insulin-producing cells of the pancreas or the multiple connected neuron in the brain.
Research on stem cells is advancing knowledge about how an organism develops from a single cell and how healthy cells replace damaged cells in adult organisms. This promising area of science is also leading scientists to investigate the possibility of cell-based therapies to treat disease, often referred to as regenerative or reparative medicine.
Scientists primarily work with two kinds of stem cells from animals and humans: embryonic stem cells and adult stem cells, which have different functions and characteristics. Scientists discovered ways to obtain or derive stem cells from early mouse embryos more than 20 years ago. Many years of detailed study of the biology of mouse stem cells led to the discovery, in 1998, of how to isolate stem cells from human embryos and grow the cells in the laboratory. The human embryos used in these studies were created for infertility purposes through in vitro fertilization procedures and when they were no longer needed for that purpose, they were donated for research with the informed consent of the donor.
Stem cells are considered to be one of the most fascinating areas of biology today.

Adapted from 'Stem cell information', the official NIH resource on stem cell research, http://stemcells.nih.gov

Transcranial magnetic stimulation

Transcranial magnetic stimulation (TMS) is the use of powerful rapidly changing magnetic fields to induce electric fields in the brain by electromagnetic induction without the need for surgery or external electrodes. TMS was originally developed as a tool in brain research, and has been used to stimulate or suppress brain activity in experiments on human subjects.
TMS is currently under study as a treatment for severe depression and auditory hallucinations. It is particularly interesting as it may provide a viable treatment for certain aspects of drug resistant mental illness, particularly as an alternative to electroconvulsive therapy. TMS is also under investigation for the treatment of drug-resistant epilepsy.
When properly combined with MRI and EEG, TMS becomes a brain mapping tool. By stimulating different points of the cortex and recording responses, one may obtain maps of functional brain areas. By measuring EEG, one may obtain information about the healthiness of the cortex (its reaction to TMS) and about area-to-area connections.

Adapted from Wikipedia,
http://en.wikipedia.org/wiki/Transcranial_magnetic_stimulation

Appendix 2

| Contributors Biographical Information |

Biographies Experts & Stakeholders

DEMOSTHENES AGRAFIOTIS, PhD started his academic career in the fields of chemical engineering and business administration and cultural sociology, but for the past 20 years has been Professor of Sociology at the Greek National School of Public Health. His research interests lie in socio-cultural dimensions of the areas of science, technology and society and the sociology of health. Professor Agrafiotis has also published volumes of poetry and essays and exhibited photography, paintings, drawings and installations in Greece and elsewhere.

JAN WILLEM BERKELBACH VAN DER SPRENKEL, PhD is Head of the Department of Neurosurgery at Utrecht University Medical Centre (UMCU). After completing a doctoral thesis on cerebral ischemia in 1988, he became fully qualified as a neurosurgeon in 1994. Before moving into his present post, he held a number of neurosurgery positions at the Utrecht Central Military Hospital and UMCU itself. He also serves on the staff of the Meander Medical Centre.

JAMES BUTCHER, PhD is a senior editor at *PLoS Medicine*, a new medical journal published by the Public Library of Science. Prior to that he was editor of *The Lancet Neurology*, a monthly review journal for clinical neurologists. The journal contains news, academic reviews, and opinion pieces on a wide range of neurological subjects. Following a doctorate in neurophysiology, he worked variously as a medical writer at Sanofi Winthrop Ltd, commissioning editor for the publication *Inside the Human Body* and, in 2000 to 2001, as a senior editor at *The Lancet*.

AXEL CLEEREMANS, PhD is Director of the Cognitive Science Research Unit at the Free University of Brussels (ULB) and a Senior Research Associate of the National Fund for Scientific Research. His research is dedicated to exploring the relationships between learning and consciousness, using behavioural, imaging, and modelling methods. Author and editor of several books dedicated to consciousness, Dr Cleeremans is currently editor of the journal *Psychologica Belgica*; he is also deputy-secretary general of the National Committee for the Psychological Sciences at the Belgian Royal Academy of Arts and Sciences, Vice-President of the Belgian Psychological Society, and an Executive Board member for the Association for the Scientific Study of Consciousness.

JOHAN DEN BOER, PhD holds the Chair of Biological Psychiatry at the State University Groningen. In addition to clinical neuroscientific studies in which neuroimaging plays an important role, he is involved in preclinical research into gender differences and molecular biological consequences of stress. His interdisciplinary interest is reflected in his recent book *Neurophilosophy: brain, consciousness and free will* (published

in 2003 in Dutch). Professor den Boer is also Distinguished Fellow of the International Society of Affective Disorders and a past President of the Interdisciplinary Society of Biological Psychiatry (1992-1997). He serves on the editorial boards of several psychiatry journals, and is the editor of a number of handbooks and academic textbooks.
In 1989, Professor den Boer was awarded the Ramaer Medal by the Dutch Society of Psychiatry for outstanding scientific research on the biological determinants of anxiety disorders.

BERIT ANDERSEN FABER, LLM is a specialist in biolaw. The Executive Director of the Danish Council of Ethics, she has participated in expert committees for the EU and the European Council in connection with the authorisation system for medicinal products and the Council of Europe's Convention on Human Rights and Biomedicine. She is co-author with Linda Nielsen, Rector of The University of Copenhagen, of *Ethical principles in European regulation of biotechnology - possibilities and pitfalls.*

SUSAN GREENFIELD, DPhiL is Professor of Pharmacology at the University of Oxford. She is also Director of the Royal Institution of Great Britain, where she heads a multi-disciplinary research group exploring novel brain mechanisms linked to neurodegeneration, and a Founding Director of BrainBoost Ltd, a company developing non-pharmaceutical approaches to Alzheimer's Disease. She has made major contributions to the communication of science to the general public. Among other television and radio work, she wrote and presented a major six-part television series on the brain and mind, *Brain Story*, in 2000. Her involvement in science policy culminated in the production of the Greenfield Report on women in science, engineering, and technology. Baroness Greenfield's work has earned her numerous honours, including being voted 'Woman of the Year' by *The Observer* in 2000, an Honorary Fellowship of the Royal College of Physicians, the 'Golden Plate Award' from the Academy of Achievement, Washington, a CBE, a non-political Life Peerage and the Ordre National de la Legion d'Honneur. Her latest book, *Tomorrow's People: How 21st Century technology is changing the way we think and feel*, explores human nature and its potential vulnerability in an age of technology.

ROLAND JOUVENT, MD is Professor of Psychiatry at Pitié-Salpétrière Hospital and Director of the Vulnerability, Adaptation and Psychopathology Laboratory at CNRS (the French National Scientific Research Centre) and University of Paris VI. His work has focused on developing a clinical and experimental approach to emotional control. He introduced the concept of the transnosographic approach, enabling models from cognitive neuroscience to be applied to psychiatry. He has also developed clinical evaluation instruments which have improved our understanding of the symptoms and treatment of depression. After discovering that certain perceptive disorders may predispose people to depression and addiction, he is now working with his team to develop therapeutic tools using ICT. Professor Jouvent was awarded the CNRS Silver Medal in 2000, and is President of the Scientific Advisory Council of the French National Laboratory for Drugs Screening.

FLAVIO KELLER, MD is Professor of Physiology and Neurophysiology, and Director of the Laboratory of Developmental Neuroscience at the University Campus Bio-Medico as well as Visiting Professor of Philosophy at the University of the Holy Cross, both in Rome. His main research field is the development and plasticity of the nervous system, with a special focus on Autistic Disorder. He has published research papers in leading scientific journals such as *Science, Journal of Neuroscience, Neuron, Journal of Biological Chemistry*, and others. More recently, he has started theoretical work on the problem of the body-mind relationship and of free will. Professor Keller has served as Scientific Coordinator of an EU research network on autism and as an expert evaluator of research proposals for the EU's Life Sciences research program of the 4th and 6th Framework Program.

DETLEF LINKE, PhD has been Professor of Clinical Neurophysiology and Neurosurgical Rehabilitation at Bonn University since 1982, and is the co-founder of the Centre for Age-Research there. He has conducted clinical research into selective anesthesia of the cerebral hemispheres for dynamic functional localisation, and his research into epilepsy has earnt him the Alfred Hauptmann Prize. Professor Linke has written 12 books concerning the interface between neurophysiology and fields such as religion and art. He is also the Vice-President of the Society for the Philosophical Study of Genocide and the Holocaust.

PETER MOLNAR, MD, PhD, Professor of Behavioural Sciences at the University of Debrecen in Hungary is a neuroscientist, psychotherapist and clinical psychologist with a particular interest in emotion, empathy and motivation. He has played an active role in incorporating medical psychology and behavioural sciences into the medical curricula in Hungary, and has also been involved in health policy issues. Professor Molnar currently serves as Lead Scientist on communication issues in a large international Leonardo da Vinci Programme aimed at achieving a systematic improvement of primary health care pediatricians' competencies in Hungary, Sweden and Greece.

RICHARD MORRIS, FMEDSCI, FRS, FRSE is Professor of Neuroscience at the University of Edinburgh. His research focuses on neurobiological mechanisms of memory. He is the life sciences co-ordinator of the Foresight Cognitive Systems Project, which has developed a vision for the future development of cognitive systems through an exploration of recent advances in neuroscience and computer science. Professor Morris holds an Adjunct Professorship of Psychology at the Norwegian University of Science and Technology, and is a Guarantor of Brain and a member of the Scientific Advisory board of the Alzheimer's Research Trust.

PAOLO LUCIO MORSELLI, MD after several years spent at the Milan Medical School and at the Medical College of Virginia, Richmond, USA, he pursued a career in private sector clinical research until his retirement in 1994. He was Chairman of the ILAE Commission on Antiepileptic drugs. He has held visiting professorships in Clinical Neuropsychopharmacology and in Psychiatry and Clinical Neuro-psychopharmacology, in Psychiatry and Clinical Psychopharmacology in France, Spain and Italy. Currently Dr Morselli is Secretary General of GAMIAN-Europe, which he founded and which is a European federation of advocacy organisations operating in the field of Mental Health. Today GAMIAN-Europe represent more than 85,000 patients suffering from mental disorders. His current interests lie in the social and therapeutic aspects of mental disorders, as well as the problems linked to the stigma associated with mental illness and the role of patient advocacy organisations in the education and in the long term management of sufferers. Dr Morselli is also charter member of the International Society for Bipolar Disorders (ISBD), where he serves on the Board as Vice President since 1999. Dr Morselli's distinctions include the Ambrogino d'Oro award, the Ambassador for Epilepsy Award, the ASPET ILAE award and the Prix Galien.

ZOLTAN NEMES, MD, PhD is currently Békésy György Professor in the Department of Biochemistry and Molecular Biology at the University of Debrecen. He is investigating the formation and composition of pathological protein aggregates encountered in Alzheimer dementia, Parkinson's disease, Huntington's chorea, several different neurodegenerative diseases and also in "normal" ageing. A particular focus of his research is the rational design of novel therapeutic approaches to rescuing neurons.

ALISTAIR NEWTON is Executive Director of the European Dystonia Federation. He has dystonia, a neurological disorder which often causes uncontrollable spasms in muscle groups, since 1984, and has been involved with patient advocacy since 1991. During his work on behalf of a number of British, European and international patient and scientific organisations (including the *British Medical Journal*, The European Federation of Neurological Societies, the European Federation of Neurological Associations, the European Brain Council and the World Health Organisation), he has made presentations at a number of conferences and has led or contributed to many projects which aim to improve the lives of people living with brain disorders .

JOCHEN OEHLER, PhD is Professor of Biology in Medicine at Dresden University of Technology. His research interests lie at the interface between neurobiology and behavioural science, and between behavioural ethology and evolutionary studies. Prior to his recent appointment as professor, he held various academic posts, firstly at Humboldt-University-Berlin, and since 1974 at Dresden.

JES OLESEN, DRMEDSCI, FRCP has been Professor of Neurology at the University of Copenhagen since 1985. He chairs a large research group there, with major projects in most areas of headache research, as well as conducting research into cerebral blood flow and aspects of multiple sclerosis. His many publications include the internationally acknowledged reference volumes, *Basic mechanisms of headache* and *The headaches*. Among other positions, he has been the President of the European Brain Council (which he founded) since 2001. He co-founded the European Federation of Neurological Associations in 2000, and has also founded and co-edited numerous scientific journals. Professor Olesen's work has been recognised with various awards, including the John J. Bonica Award in 1996.

IAN RAGAN, PhD is Executive Director at the Lilly Research Centre of Eli Lilly & Co. Ltd., where he is responsible for the company's European neuroscience research programme devoted to the discovery of innovative treatments for psychological and neurological illnesses. In addition, he is Executive Director of European Scientific Affairs and acts as Lilly's spokesperson on R&D matters within the EU. In this capacity he is currently involved with the EC Framework Programme initiatives to increase pharmaceutical R&D investment in Europe for the benefit of patients. He is a member of the EFPIA Research Directors Group, the European Brain Council and Chair of the R&D Committee of the Association of British Pharmaceutical Industries. Before joining Eli Lilly in 2000, he spent 14 years at Merck, Sharp and Dohme Research Laboratories. Dr Ragan's previous academic career culminated in a senior lectureship in biochemistry at the University of Southampton. His interests cover the whole area of neuropharmacology and he is the author of over 150 publications

BERNARD REBER, PhD is a moral and political philosopher and Researcher at the CNRS (National Centre for Scientific Research) at University of Paris V, where he is a member of CERSES, the Meaning, Ethics and Society Research Unit. He also teaches university courses on public scientific controversies and the precautionary principle. Dr Reber's research interests are in applied ethics, participatory technological assessment and the ethics of technology.

ANDREAS ROEPSTORFF, MSC, MA, PhD is Associate Professor at the Centre for Functionally Integrative Neuroscience & Department of Social Anthropology at the University of Aarhus. As an anthropologist in neuroscience, he seeks to maintain a dual perspective: he studies the workings of the brain, particularly at the levels of consciousness, cognition and communication, but is also interested in how brain imaging, as a field of knowledge production, relates to other scientific and public fields.

STEVEN ROSE, PhD became Professor of Biology at Britain's Open University in 1969. There, he established the Brain and Behaviour Research Group, which he has directed ever since. His research on the

cellular and molecular mechanisms of learning and memory has led to the publication of some 300 papers and brought him various international honours. Among the 15 books to his credit, *The Chemistry of Life* became a minor classic over five editions from 1964 to 1999, while *The Making of Memory* was winner of the 1993 Rhone-Poulenc/Royal Society Science Book Prize. He is also interested in science-related social issues. He helped establish the British Society for Social Responsibility in Science and was co-convenor of the Science Engineering and Technology Policy Forum, set up to advise the Labour party before the 1997 election. Professor Rose is a well-known radio and television figure.
His new book, *The 21st Century Brain* will be published early in 2005.

WRYE SENTENTIA, PhD is director of the Center for Cognitive Liberty & Ethics (CCLE), a nonprofit research, policy and public education centre that focuses on the intersection of emerging neurotechnologies with human freedom. For the past two years, Dr Sententia and the CCLE have been sponsors of the National Science Foundation's initiatives aimed at "Converging Technologies for Improving Human Performance." In 2002, she advised the President's Council on Bioethics, in Washington D.C., on enhancement technologies and drugs. She is currently working with the California Legislatures' Joint Committee on 'Preparing California for the 21st Century'. Dr Sententia recently completed a doctoral dissertation at the University of California Davis on the impact of new media and biotechnology on the social ethics of privacy, autonomy and choice.

PAOLO VALERIO, MD is Professor of Clinical Psychology at the Naples University Federico 2nd Medical School. Among other positions, he is also Director of the Postgraduate Course in Clinical Psychology, Head of the Clinical Psychology Department at Naples University Hospital, Co-ordinator of the Counselling Service at the Italian Military Air Force Academy and Co-ordinator of the Counselling Service for University Students. His research interests are in psychodynamic counselling, gender identity and emotional disorders in late adolescents and young adults.

CHRISTINE VAN BROECKHOVEN, PhD, DSc is a full professor in the Department of Biomedical Sciences at the University of Antwerp and scientific director of the Department of Molecular Genetics of the Flanders Interuniversity Institute for Biotechnology. She is also a research director at the Born Bunge Foundation. Her team specialises in the molecular genetics of complex genetic diseases of the central and peripheral nervous system. Professor Van Broeckhoven has been awarded several scientific prizes, including the 1993 Potamkin Prize for her contribution to the identification of APP as a gene for Alzheimer's disease. She is a member of the Royal Flemish Academy of Sciences and the Arts of Belgium.

MARK VAN BUCHEM, MD, PhD is Professor of Neuroradiology and Head of the Neuroradiology Section at Leiden University Medical Centre. He is also attending radiologist at the Department of Radiology there. In addition to numerous other professional activities with the International Society for Magnetic Resonance in Medicine and other organisations, he is currently President of the Dutch Society of Neuroradiology. Professor Van Buchem is also involved in international collaboration projects with the University of Pennsylvania Medical Center (image post-processing methodology; analysis of volumetric MTI data) and the British Institute of Neurology (analysis of volumetric MTI and MR-diffusion data; voxel-based morphometry analysis).

Appendix 3

| The Partner Consortium ECD |

King Baudouin Foundation

The King Baudouin Foundation is a public benefit foundation founded in 1976, 25 years after King Baudouin's ascension to the throne. The KBF is independent of government support or influence yet adopts a pluralistic approach to its work to improve living conditions for the population. A European foundation based in Belgium, the KBF is active at local, regional, federal, European and international levels, capitalising on its location in Brussels, the capital of Europe, of Belgium and of Belgium's two large Communities. The KBF is a founding member of the European Foundation Centre (EFC) and the Network of European Foundations for Innovative Cooperation (NEF). Its core programmes over the next few years include: Social Justice, the Civil Society, Governance, and Funds and Contemporary Philanthropy. The Governance Programme aims 'to increase the impact of civil society on the debate and the decision-making process regarding science and technology as well as modes of production and consumption'. The KBF aspires to create a stimulating environment where individuals and institutions have access to new forms of social debate, decision-making processes and political approaches. This will be achieved by giving citizens a key role and encouraging them to become involved in such issues based on their own questions, concerns and capabilities. In 2003, the KBF organized the first Citizens' Conference in Belgium on the issue of genetic testing.

More information: www.kbs-frb.be

ECD members
- Gerrit Rauws, Director
- Tinne Vandensande, Project officer
- Ann Nicoletti, Assistant

Rathenau Institute

The Rathenau Institute is the Dutch national technology assessment organisation. It is an independent organisation established (in 1986) and financed by the Dutch Ministry of Education, Culture and Science. The Rathenau Institute's task is to support social debate and political opinion formation on issues resulting from or linked to scientific and technological developments, including ethical questions.. To that end, it promotes research, organises debates and hearings and issues publications. Its main client is the Dutch Parliament.

More info: www.rathenau.nl.

ECD members:
- Jan Staman, Director
- Rinie van Est, Assistant Project Director
- Sara Heesterbeek, Project manager

University of Westminster, Centre for the Study of Democracy

The Centre for the Study of Democracy is a leading postgraduate and research centre and is well known for its interdisciplinary work, and comprises a team of more than a dozen internationally recognised scholars, whose teaching and research concentrates on the interplay of democracy, states, cultures, technology and civil societies. The CSD's research and teaching activities are guided by clear scholarly aims and a distinctive commitment to:

- Conducting open and critical enquiries into politics, governance, international relations, media, cultural and technology studies and other related disciplines;
- Synthesising the study of democracy, politics, international studies, civil society, technology and culture;
- Pursuing comparative research in local, regional and global contexts;
- Fostering awareness of the advantages and disadvantages of democracy at the local, national and international levels.

Scientific and methodological advice as well as social-scientific expertise will be sourced from the 'Participatory Technology Assessment' and Science & Technology Studies team based at the CSD at the University of Westminster.

More information : www.wmin.ac.uk/csd

ECD members

- Simon Joss, Director of the CSD
- Alison Mohr, Research Fellow

Flemish Institute of Science and Technology Assessment

viWTA, the Technology Assessment Institute of the Regional Parliament of Flanders, with SPIRAL of the University of Liège for the French Community, bear shared responsibility for and represent the Belgian input in the project. The Flemish Regional Parliament and Government are responsible for legislation and policy in education, environment, and science and technology policy, though some competences are shared at Belgian federal level. In Flanders, the emergence of parliamentary TA as a multidisciplinary practice occurred a few years after it happened in neighbouring countries. Previously, TA was conducted mainly by autonomous semi-governmental institutes and at universities. However, it became increasingly difficult to ignore the growingly intensive social and public debate on waste disposal, genetic engineering and food safety (dioxins). The latter became a key issue in the 1999 general elections and in November 1999, a proposal for a TA institution was submitted to the Flemish Parliament and discussed in Joint Committees. viWTA is first and foremost a parliamentary TA institution. viWTA's research areas include both foresight studies and upstream TA as well as analysis of current technological developments by promoting public debate. This provides a consistent stream of information on the interaction between society and technology, to the benefit of Parliament, interest groups and the general public. viWTA has an advisory role in which it addresses questions from the Flemish Parliament on science and technology policy.

More info: www.viwta.be

ECD members:
- Robby Berloznik, Director of the viWTA
- Stef Steyaert, Senior Researcher
- Willy Weyns, Senior Researcher

Danish Board of Technology

The Danish Board of Technology is the parliamentary technology assessment institution of Denmark. The role of the Board is to perform technology assessment, advice the parliament and the government, and to further debate the issue of technology in Denmark. The methodologies used include traditional expert-based analysis, parliamentary debates, and participatory methods involving stakeholders and lay people, as well as public debates and enlightenment activities. The Danish Board of Technology was brought into being in order to disseminate knowledge about technology, its possibilities and its effects on people, on society and on the environment. The Board aims to promote the ongoing discussion of technology, to evaluate technology and to advise the Danish Parliament (the Folketing) and other governmental bodies in matters pertaining to technology. The Danish Board of Technology is an independent body established by the Danish Parliament in 1995 and is the successor of the Technology Board, which was set up as a statutory body in 1986. The Parliament's Research Committee is the Board's ongoing liaison with the Parliament. Once a year an annual report is submitted to the Parliament and the government.

More info: www.tekno.dk

ECD members:
- Ida-Elisabeth Andersen, Project manager
- Ida Leisner, Communication officer

Cité des Sciences et de l'Industrie

Located in north-eastern Paris, the Cité des sciences et de l'industrie (CSI) (established in 1986) houses several different activities which all aim to achieve "public understanding of science", including permanent and temporary exhibitions, covering an area of 30,000 m^2, visited by approximately 1.5 million visitors p.a.; an area dedicated to children, the "Cité des enfants"; a reference library, the "Médiathèque", which attracts approximately one million readers per year; a congress centre, and several auditoriums dedicated to film and lecture programs. The *Cité des sciences* works in close collaboration with others science centres, including three in Paris (Musée d'Histoire Naturelle, Palais de la Découverte, Conservatoire des Arts et Métiers) , the network of the "Centres de Culture scientifique et technique", and with 30 centres in the main cities in France. These partners bring a national dimension to any collaborative initiatives.

More info: www.cite-sciences.fr

ECD members:
- Roland Schaer, Director Science & Society
- Marie-Agnès Bernardis, Project manager

Stiftung Deutsches Hygiene-Museum

The *German Hygiene Museum* was opened in 1930. Now, as a modern science museum, its task is to reflect the impact of Life Sciences on society in the 21st century. It is a "Universal Museum of Man" presenting the human body and human health in its new permanent exhibition. Temporary exhibitions like "Adventures into Mind: The brain and Thought" (2000), "The Imperfect Human Being" (2001-02), "Man and Beast. A Paradoxical Relationship"(2002-03), and "The Ten Commandments" (2004) focus on mankind and its interaction with society and the environment. Events such as lectures, discussions, performances and congresses on the topics of the exhibitions provide an independent public forum for a deeper dialogue between science and society. The museum's educational task is to promote discussion on the trends in science and research. The museum organised the first nationwide Citizens' Conference on Genetic Testing in 2001

The museum's current activities include travelling exhibitions, publications, educational programs for students, advanced training for teachers and a laboratory for students to carry out experiments in the field of molecular biology. A children's museum is planed for 2005.

More info: www.dhmd.de

ECD member:
Jörg Naumann, Head of the Science Unit

Città della Scienza

The *Fondazione Idis-Città della Scienza* is a socially orientated, non-profit making organisation for developing science, culture and innovation to overcome the serious economic difficulties in the south of Italy. *Città della Scienza* is a structure of 65,000 m^2, of which 27,000 m^2 is indoor space. It fulfils different functions and offers services including a Science Centre, a Business Innovation Centre (BIC), a Centre for Advanced Training and Vocational Guidance and a Congress Centre. The *Science Centre*, whose primary aim is to establish a dialogue between science and society, has 8,000 m^2 of exhibition spaces and functions to increase awareness of science and technology to the wider public. The *Science Centre* promotes many events, such as the important Annual Science Festival, also called 'Futuro Remoto'. It is a 'forum', where researchers and scientists can meet the wider public, providing the latter with access to scientific information and stimulating debate on contemporary topics. The Business Innovation Centre, BIC, comprises supporting services for economic innovation. *Città della Scienza* is an important reference point for applying the participative methodology, EASW, European Awareness Workshop, promoted by DG Enterprises of the European Commission and cited by the Ministry for the Environment, ANPA and ENEA, as good practice for the promotion of social participation in projects whose aim is to improve the environment and to stimulate sustainable development.

More info: www.cittadellascienza.it

ECD members:
- Luigi Amodio, Director
- Jennifer Palumbo, Science center department
- Alessandra Drioli, Director's Assistant
- Carlo Guardascione, Communication officer

Science Museum (SM)

The Science Museum in London has a long tradition of liaising with its visitors - and with the wider public - on issues of contemporary importance. The Museum, which now has 3 million visitors per year, has become a national focus of public understanding of science activities, for example the first UK consensus conference on plant biotechnology in 1994, and the opening in 2000 of the Wellcome Wing, the world's first Museum development concentrating exclusively on contemporary science and technology. Now, with the opening of the Museum's new Wellcome Wolfson Building in November 2003, the Museum will be ideally placed to organise and to host a national debate on scientific issues of public interest. The building will include the world's first purpose-built centre for publicscience events about contemporary issues: in addition to meeting rooms, a TV studio, a cafe and a host of other public facilities, the entire building will be 'wired' so that activities that take place in the Centre can be linked to other centres in the UK and to everyone on the internet. The Building will also accommodate 75 staff who are working on the public dimensions of science.

More info: www.sciencemuseum.org.uk

ECD members:
- Graham Farmelo, Senior Research Fellow at the Science Museum
- Lisa Jamieson, Head of Programming at the Dana Centre at the Science Museum
- Tom Ziessen, Assistant at the Dana Centre at the Science Museum

Medical and Health Sciences Centre, University of Debrecen

The University of Debrecen and its Medical and Health Sciences Centre represent one of the focal points of medical research in Hungary. Several of its basic science and clinical departments carry out brain research, some related to clinical genetic and genomic research. In clinical genetics there is a particular focus on mutations effecting coagulation, lipid metabolism and certain cases of tumorigenesis (Department of Clinical Biochemistry and Molecular Pathology), cytogenetics of childhood leukemias (Department of Pediatrics) and neurodegenerative disorders such as Alzheimer's disease, dementia, etc. (Department of Anatomy, Neurology). In 2001 the university established the Debrecen Clinical Genomics Centre to provide a technological base and core laboratories for biobanking, genome wide gene expression studies and bioinformatics at UD and in the region. The Genomic Centre carries out not only collaborative studies with UD laboratories but also contract research with Hungarian SMEs and large pharmaceutical companies (i.e. Pfizer Global Research).

More info: http://www.ceebd.co.uk/ceeed/un/hu/hu005.htm

ECD members:
- Peter Molnar, head of the Department of Behavioural Sciences at the University of Debrecen
- Marta Csabai, social- and clinical psychologist, Department of Behavioural Sciences
- Laszlo Fesus, president of the Medical and Health Sciences Centre

Eugenides Foundation

Established in 1956 by Eugene Eugenides, the Eugenides Foundation is an independent, non-profit welfare organisation whose mission is to enhance scientific, technological and technical education of young people in Greece and to promote Science and Technology to the wider public. Over a period of 50 years, the EF has contributed substantially to public understanding of science and technology in Greece. Through the Planetarium, publications, the library, scholarships and other activities, including special exhibitions and lectures, the EF offers a combination of education and entertainment to many visitors, particularly young students, a great number of whom were inspired to pursue successful academic careers in natural sciences and engineering. The EF has also actively supported their academic development through a series of scholarships and grants. The EF aspires to play the role of a multifaceted educational and technological hub unique in Greece, where students, educators and the wider public will actively participate through learning and exploring. Today, the EF has a new vision for the new century: to contribute to the development of an environment which would allow for a more balanced flow of information to the citizen on S & T issues. For this purpose, the EF is currently developing a Research Institute for Science - Society Interfaces, which will closely cooperate with all relevant stakeholders in Greece and abroad.

More info: www.eugenfound.edu.gr

ECD member:
- Glykeria Anyfandi, Project development

Université de Liège, SPIRAL

SPIRAL, the Scientific and Public Involvement in Risk Allocations Laboratory of the University of Liège, together with viWTA for the Flemish Community, bear shared responsibility for and embody the Belgian input in the project. Since 1995, the SPIRAL team has conducted fundamental and applied research in the fields of contemporary risk management, environmental and siting conflicts and public decision-making. The approach used is clearly interdisciplinary, associating researchers from different backgrounds: social and political sciences, economics, communication and anthropology, together with specialists from relevant technologies. The group promotes a European approach in its work in order to compare the evolution of Belgian federal and regional policies with those of neighbouring countries. Research conducted in the last five years by the SPIRAL mainly concerns analysis and evaluation of public policy in the following areas : risk management, crisis management and public participation in policy-making procedures. Research at SPIRAL mainly focuses on the new production conditions of adhesion mechanisms and social acceptability of public policies. This includes policies linked to these areas, including communications policy, the development of individual preferences in matters of acceptability, environmental conflict resolution, institutional expertise, and so on.

More info: www.ulg.ac.be/spiral/

ECD members:

- Catherine Zwetkoff, Professor at the University of Liège, Director of spiral
- Sebastien Brunet, Professor-Assistant at the University of Liège.

Index

Index of names

Index of words